David Attenborough is one of the world's leading naturalists and broadcasters. His distinguished career spans more than sixty years, and his extraordinary contribution to natural history broadcasting and film-making has brought him international recognition, from *Life on Earth* to *Frozen Planet*, *Planet Earth* to *The Blue Planet*, and He has achieved many professiona merits, including the CH and OM.

T0109655

Praise for *Life on Earth* (2018)

'It does not disappoint. The new *Life on Earth* is as glorious as the first, if not more so for the sole reason that it has been considerably updated'   Adam Rutherford in the *Guardian*

'A beautiful and wide ranging work. The breadth of natural history covered is extraordinary and mesmerising. *Life on Earth* is still breathtakingly rich, and we would know far less about it were it not for Attenborough's wonderful skills of communication over the years: our cultural and scientific lives would be poorer without him'   *New Scientist*

Praise for David Attenborough

'An elegant and gently funny writer'   *The Times*

'His writing is as impressive and as enjoyable as his TV programmes and there can be no higher praise'

Daily Express

'A great educator as well as a great naturalist' Barack Obama

'When I was a young boy I used to love turning on the television and watching David's programmes and really feeling like I was either back out in Africa or I was learning about something magical and almost out of this planet'

Prince William

'Sir David is a wizard of television, and, like Gandalf or Dumbledore, he has a near-magical gift for combining warmth and gravitas . . . the man who, for me, exemplifies the best in British broadcasting'   Louis Theroux

OTHER TITLES BY DAVID ATTENBOROUGH

The Life Trilogy

Life on Earth (new edition 2018)

Living Planet (new edition 2021)

The Trials of Life (new edition 2022)

The Life on Land Series

The Private Life of Plants (1994)

Life in the Undergrowth (2005)

Life in Cold Blood (2007)

The Life of Birds (1998)

The Life of Mammals (2002)

Other Titles

Zoo Quest to Guiana (1956)

Zoo Quest for a Dragon (1957)

Zoo Quest to Paraguay (1959)

Quest in Paradise (1960)

Quest under Capricorn (1963)

The Tribal Eye (1976)

Life on Air: The Memoirs of a Broadcaster (1987)

David Attenborough's Life Stories (2009)

David Attenborough's New Life Stories (2011)

Drawn from Paradise: The Discovery, Art and Natural History of the Birds of Paradise (2012) with Errol Fuller

Adventures of a Young Naturalist: The Zoo Quest Expeditions (2017)

Journeys to the Other Side of the World: Further Adventures of a Young Naturalist (2018)

A Life on Our Planet: My Witness Statement and a Vision for the Future (2020) with Jonnie Hughes

The
Trials of Life

A Natural History of
Animal Behaviour

David
Attenborough

WILLIAM
COLLINS

William Collins
An imprint of HarperCollins*Publishers*
1 London Bridge Street
London SE1 9GF

WilliamCollinsBooks.com

HarperCollins*Publishers*
Macken House, 39/40 Mayor Street Upper
Dublin 1, D01 C9W8, Ireland

First published in Great Britain by William Collins Sons & Co. Ltd.
and BBC Books: a division of BBC Enterprises Ltd. in 1990
Fully updated and republished by William Collins in 2022
This William Collins paperback edition published in 2023

3

A catalogue record for this book is available from the British Library

ISBN 978-0-00-847790-5

Designed and typeset in Baskerville by Tom Cabot/ketchup
Picture research and editing by Rachelle Morris/Nature Picture Library

Printed and bound in the UK using 100% renewable
electricity at CPI Group (UK) Ltd

MIX
Paper | Supporting
responsible forestry
FSC™ C007454
FSC
www.fsc.org

This book is produced from independently certified
FSC™ paper to ensure responsible forest management.

For more information visit: www.harpercollins.co.uk/green

CONTENTS

Introduction

This book, and the television series that was filmed at the same time as it was being written, is the last in a trilogy of natural histories. The first was *Life on Earth*. That set out to examine the vast diversity of animal life. Why should there be such an extraordinary variety of animals all doing somewhat similar things? Why should a whale have warm blood and lungs, whereas a similarly-sized swimming monster, the whale shark, has cold blood and gills? Why do indigenous Australian mammals rear their young in a pouch whereas mammals in the northern hemisphere retain their offspring within a womb and nourish them by means of a placenta? The answers to such questions can only come from an understanding of history. So *Life on Earth* traced the development of animal life from its beginnings some three thousand million years ago until today and illustrated its great episodes by taking living animals as examples.

The second book, *Living Planet*, concentrated on the other great influence that shapes the bodies of animals, the environment. Mammals that live in deserts tend to have longer ears and legs than their equivalents in cooler areas – bodies shaped in that way are more efficient at losing heat; land birds, marooned on remote islands, tend to become flightless – with no land predators present, they have no need to take to the air. Other organisms with which an animal shares its environment also have their influence. So the fur of an Arctic hare that keeps it warm in winter also turns white when the snow comes in order that its wearer shall remain concealed from predators. *Living Planet* surveyed ecological communities throughout the world, ranging from the baking deserts to the humid rainforest, from the depths of the ocean to the highest layers of the atmosphere.

Thus, the first two books were concerned with the bodies of animals and the way they have been shaped. This last book looks at the way animals use those bodies, the way they behave.

Behaviour is perhaps the most obviously exciting aspect of natural history. It is full of action and drama – a killer whale surging up a beach to grab a young sea-lion; an ant navigating across a Saharan sand dune by taking repeated observations of the sun; a mother bat fighting through crowds of begging infants on the roof of a cave in order to give her milk to her own baby and no other. Animal behaviour might, therefore, have been the obvious choice for the first subject in this trilogy. The fact is, however, that ten years ago when my colleagues and I began work on the television series, we could not have witnessed many of the actions that I can now describe and certainly could not have recorded pictures of them.

The reasons for this are partly technical. During the decade before this book was written, there had been major advances in both film technology and electronics. As a consequence we were just beginning to watch and record events in light so dim that only recently such events were beyond the sensitivity of any photographic emulsion or even our eyes. Now, with the aid of electronic image-intensifiers and super-sensitive film, we can record fire-flies flashing synchronously like lavish Christmas illuminations in the mangrove swamps of Malaysia and see them not just as featureless spots of light but as tiny beetles engaged in elaborate courtship rituals. Now, with fibre-optic probes developed for use in medicine, we can see for the very first time what happens within the huge globe formed by a million army ants bivouacked beneath a log in the rainforest of Panama.

But more important than such technical advances has been the great increase in the number of scientists actively involved in observing animals in the wild. Almost every group of large animals is now being studied by scientists somewhere. These researchers have become so knowledgeable about their subjects and understand them so intimately that they have been able to guide us to the right place at the right time in order to see exactly that aspect of behaviour that was of particular interest to us.

Few scientific disciplines demand greater dedication or the endurance of such harsh physical circumstances as studying wild animals in their natural environments. Christophe Boesch is a Swiss zoologist who worked with chimpanzees that live in the thick forest of the Ivory Coast in West Africa. He and his wife Hedwige started their project ten years before we visited them, spending every day for weeks on end walking quietly through the forest. For the first year or so he counted it a good day if he got a brief glimpse of a chimpanzee. He did not allow himself to bribe the apes with food and so lure them out in the open towards him, believing that to do so would distort their natural behaviour and so invalidate his findings. Only after four years of unrelenting observation and tracking did the chimpanzees become sufficiently accustomed to their silent human shadow to have no fear of him and ignore him.

Several more years passed before he was able to recognise with certainty the different individuals in the group, as was necessary for his work. Eventually he came to know every one of the sixty or so chimpanzees in the group by sight, but he could also recognise most by their voices, even when they called from a considerable distance. Every day he followed them as they travelled through the forest, stopping when they paused to feed, running when they started to travel at speed. Only after they started making their beds in the tree tops each evening did he leave them. And in the morning, before the sun was up, he set off from his house in the forest to rejoin them, if necessary running for an hour or so to make contact with them again before they moved away to some part of the forest where they might be difficult to find. The result of all this persistent and punishing work was to reveal among many other things that forest-living chimpanzees are regular hunters and have developed techniques of working in teams to catch their prey that are more elaborate than those used by any other animal except human beings. That Christophe should have allowed us, with cameras, to accompany him and film them hunting in this way was an extraordinary privilege.

He was only one of the many scientists who helped us. A list of others who helped me personally appears at the end of this book. In addition to these, a great number more most generously gave advice and practical guidance to the directors and cameramen working on the television series and so made it possible for us all to share sights that they themselves had only been able to witness after years of intensive and patient study. Our debt to them all is unpayable.

4

The science of animal behaviour, which such researchers serve, is known as ethology. I have not attempted in this book to describe the body of theory that it has developed any more than I examined theories about the mechanisms of evolution in *Life on Earth*. The reader who wants a full exposition of such subjects as selfish genes, game theory, altruism or the relationship between learning and instinct, must look to more technical texts. My concern here is to describe the happenings, rather than the psychological and evolutionary mechanisms that produce them. That, I now realise even more vividly than when I started on this project, is a big enough task.

It is not always possible to disentangle behaviour from anatomy, and to that extent there is inevitably some overlap between what I have written in this book and its two predecessors. On a few occasions a species has been described for a second time because its behaviour is unique and so extraordinary that this survey would have been inexcusably incomplete without it. But the variety of animal life is so vast that for the most part it has been possible to find different examples to make my points and if this has meant neglecting more famous and obvious instances, then that perhaps is to be welcomed.

As before, I have not encumbered the text with scientific Latinised names where there is a reasonably accurate English equivalent. This inevitably leads to some loss of precision, but those readers who wish to know exactly which animal I am describing can discover by looking up the English name in the index, where its genus if not its species will be found in italics.

All organisms are ultimately concerned to pass on their genes to the next generation. That, it would seem to a dispassionate and clinical observer, is the prime objective of their existence. In the course of achieving it, they must face a whole succession of problems as they go through their lives. These problems are fundamentally the same whether the animals are spiders or squirrels, mice or monkeys, llamas or lobsters. The solutions developed by different species are hugely varied and often astounding. But they are all the more comprehensible and engaging for they are the trials that we also face ourselves.

ONE

Arriving

It is midnight on the coast of Christmas Island in the Indian Ocean, five hundred kilometres south of Java. The November moon is in its third quarter and the tide is coming in. Behind the narrow sandy beach stands a sheer cliff of coral rock, twenty metres high. On its vertical face, clinging beneath overhangs, jammed three or four deep into cracks, are a million scarlet crabs. In places, they are so crowded that their bodies touch and the cliff seems to have been painted crimson. These crabs are found nowhere else in the world. They are large animals with glossy rounded shells eight centimetres across. All are females, each with a huge mass of brown eggs bulging beneath the semi-circular flap on her underside. They are about to spawn.

A month ago they, together with the males, left the burrows on the floor of the forest inland where they had spent most of the year and began a long march to the coast. Then the vast size of their population became dramatically apparent. There were about forty-five million of them. They moved mostly in the

7

early morning or the evening, for they dry out easily and cannot withstand the full tropical sun. But when the sun went behind clouds, and particularly after a rain shower when the undergrowth was moist, they travelled during much of the day – up to twelve hours at a stretch, compared to only ten minutes during the dry season. Nothing deterred them. In places their traditional routes cross roads made by the people who now live on Christmas Island. Thousands of the marchers were inevitably crushed beneath the wheels of the traffic but still, day after day for two weeks or so, they kept coming. When they reached the coast, the males excavated burrows and there mated with the females. The males then returned inland, but the females had to wait in the burrows for a further two weeks while their fertilised eggs matured.

And now the moment to release the eggs has arrived. The crabs have climbed down the cliffs, for their eggs must be deposited directly into the sea if they are to hatch. But this is not without hazard. Although the crabs' distant ancestors came from the sea, these are land crabs. They breathe air and they cannot swim. If they lose their hold on the rock or are swept away by the waves, they will assuredly drown.

As the tide reaches its height, the width of the beach is reduced to a few metres. The females move down from the cliffs, across the shingle to the breakers, scrambling over one another in their anxiety to get to the water. Soon the sea is fringed with a moving scarlet carpet of glinting shells, grappling legs and craning stick-like eyes. When at last the waves sluice over them, each shakes her body convulsively so that the brown eggs swill away in the water and, with a touching gesture of apparent exultation, lifts her claws above her head as if waving a salute.

At either end of the beach, where the sea beats directly on the face of the cliffs, the crabs have a harder time of it. So great is

8

the traffic between those striving to clamber down to the sea and those who, having spawned, are attempting to get back again, that many cannot reach the water. They are thus compelled to release their eggs while they are still high on the rock and a brown rain of spawn falls sporadically from as high as six metres. In the confusion, many crabs lose their foothold, tumble into the water and are swept away.

Each one of these females sheds about a hundred thousand eggs. The waves and the water beyond have become a thick brown soup. As the sky lightens in the east, the crabs leave the water's edge and are on their way back to the forest. Only a few stragglers remain on the shore. Here and there, limp bodies float in the shallows and great expanses of the beach are covered with a layer of brown grains that are not sand but eggs. The extraordinary laying is over for another year and the crabs' progeny, abandoned, must now look after themselves.

Huge numbers of the hatchlings are immediately eaten by the fish that swim in shoals around the reefs. Moray eels squirm right to the water's edge and greedily gulp down the feast. As the survivors are swept out to sea, so the larger fish, trawling with open jaws, sieve them from the water. They are helpless, drifting wherever the currents and tides take them. They feed by collecting tiny particles from the water. Periodically, they moult their thin transparent skins, changing shape as they do so. But they cannot assume their final adult form and breed unless they reach land. The vast majority of them never do so. They die unmated and without progeny. Most years the entire spawning is lost. But then, about one year in six, some fortunate swirl in the currents brings them back to the island where they first fell into the water a month earlier and at a high tide in December, a horde of tiny crablets no bigger than ants suddenly emerges from the waves and marches valiantly up the beach and on inland to restock the forest. Even then, their ordeal is not over – invasive ants, inadvertently introduced to the island in the 1930s, have eaten tens of millions of crabs in the intervening years.

The land-crabs' breeding strategy is extravagant and wasteful but successful. The multitudinous hazards that face their young – the predatory fish, the vagaries of the currents, the absence of islands over vast areas of the surrounding ocean – are met and ultimately defeated by sheer numbers. But the cost is stupendous. The average female lives for about ten years and produces in all about a million eggs. Almost all the hatchlings will die within a few weeks. But if only two of this million reach adulthood, one for each parent, then the land-crab population of Christmas Island will be maintained.

This profligate recipe for survival is used by many animals of many kinds. A single female cod can produce six million eggs in one spawning. On land, insects use the same strategy. A female fruit fly, simply because of her tiny size, can hardly be expected to produce eggs in numbers to rival a cod, but even so, she can lay two thousand in a season in batches of a hundred at a time. The really big egg-producers, however, are some of the simpler animals that live in the sea, such as corals, jellyfish, sea-urchins and molluscs. And champion of them all, whether on land or in the sea, is almost certainly the giant clam. That can produce five hundred million eggs in one gargantuan splurge. And it may perform this stupendous reproductive feat annually for up to a century.

There is, however, an alternative to this extravagance. A female, instead of manufacturing the maximum number of eggs that can be created from her bodily reserves, may produce rather fewer but give each one a better chance of survival by supplying it with food in some way, so that it is sustained during its first difficult days. Some animals place this food within the egg as yolk. In simpler creatures, granules of it are distributed evenly throughout the egg – in a frog's egg, it is concentrated at one end – and in a bird's egg it initially fills the greater part of the shell. So generous is this bequest by birds to their young that a chick needs no additional food from which to build the flesh and bones and feathers of its infant body, and it still has enough energy left over to break its way out of the shell.

Insect eggs, however, contain very little yolk. Instead, the females may help their young by placing their eggs where the minute hatchlings will find food just as soon as their heads emerge from the egg capsule. A butterfly sticks them on the leaves of the particular plant that her caterpillars eat; a blowfly on the dead flesh that her maggots will relish; and some wasps, for the sake of their young, become body-snatchers.

The Ammophila wasp, when breeding time comes, starts by digging burrows. It lives on every continent except Antarctica. A favourite site is a bare patch of earth where the surface has been baked into a crust by the harsh sun. She breaks through it by using her head like a pneumatic drill, pressing her hard sharp jaws on to the soil and vibrating it by trembling her wing muscles. Once through the crust, tunnelling is easier and she brings out loads of sand, clutched between her forelegs. When the tunnel is finished she scours the bushes and fields nearby for caterpillars.

As soon as she finds one she paralyses it, using her long sting as though it were a hypodermic syringe loaded with anaesthetic.

Then she flies back to her burrow, carrying the immobilised caterpillar beneath her. Laboriously, she drags it down into the tunnel and there, in the dark, she lays a single egg on the inert body. One burrow may eventually contain as many as half a dozen of these paralysed prisoners, each doomed in due course to be eaten alive by the wasp grub that hatches upon it. When the hole is fully stocked, the Ammophila seals it with a plug of sand made firm and smooth by hammering it with a grain of gravel held in her jaws.

Several thousand species of wasp around the world provide for their young in this way. Oxybelus, rather smaller than Ammophila, supplies her young with flies. Having seized and anaesthetised one, she does not withdraw her sting but flies back to her burrow with the fly still impaled behind her like a sausage on a stick. Pepsis, a giant among wasps with a ten-centimetre wingspan, lives in South America and grapples with bird-eating spiders as big as a human hand. After paralysing them, she amputates their legs to make the job of transport a little easier.

The burrows of these robber wasps are usually concealed so skilfully that few other animals are able to find them and rob them. But eggs, particularly those with large stores of rich yolk within them, are excellent eating, and many other animals steal them if they can. This can occur even when the egg is hidden. Many wasps and flies are parasitoids, laying their eggs inside another arthropod − generally a caterpillar − and then slowly eating it from the inside while it remains alive, until the time has come to pupate, at which point the larvae do so, and the unfortunate host generally dies. But there are some parasitoid wasps whose prey is not the primary host, but instead the parasitoid larva within it. These hyperparasitoids detect the parasitised victim by its particular odour, and then lay their eggs inside the parasitoid larva within. Both the host, and the parasitoid, eventually succumb to the voracious hyperparasitoid larva.

So precious are eggs that many parents invest a great deal of time and energy in protecting them. Several species of birds – caciques and oropendolas in South America, for example, and weaverbirds in Africa – habitually build their nests close to those of ferocious wasps which many animals take care to avoid disturbing. Oddly, the wasps pay no regard to the building birds, but they will attack any other creatures that dare to approach either their nests or those of the birds.

A Mexican fly, Ululodes, lays her eggs in batches on the underside of twigs. Having finished, she descends a little way down the twig and then lays another batch. But these are different from the first. They will never hatch. They are a little smaller, club-shaped and covered with a shiny brown fluid which neither hardens nor evaporates, but remains liquid for the three or four weeks it takes the eggs higher up the twig to hatch. If an ant, searching the twig for food, so much as touches the barrier of infertile eggs with its antennae, it recoils violently and may even lose its footing and fall. For a minute or more it cleans itself frantically. Only then does it run off to look for other food – elsewhere.

Most reptiles, having buried or concealed their eggs in some way, abandon them, but a few stay beside them and will valiantly defend them against robbers. King cobras curl around their pile of eggs, encircling it with their coils, and crocodiles stay alongside their nest of decaying vegetation for the two months or so that it takes the eggs within to hatch.

Birds, which many scientists consider to be a special kind of reptile, have no alternative in this matter. They have warm blood and so do their young inside the eggs. If the eggs are allowed to cool, once they have started to develop, the chicks within will die. So usually one or other of the parent birds must stay with the eggs for most of the time. They warm them by pressing them against brood patches, areas of skin naked of feathers which a bird may develop specially for the breeding season or have permanently on its breast concealed by the long feathers growing around them.

Chilling is the commoner danger, but in deserts there may be a risk of over-heating and that can also be lethal. So the blacksmith plover in the savannahs of East Africa will stand over its eggs, shadowing them with outstretched wings, to allow what wind there is to blow over them; and in Australia, a jabiru stork will collect water in its beak to spray over the eggs if they get dangerously warm.

The megapodes, a family of birds that lives in Australia and the western Pacific, have developed particularly ingenious techniques of incubation. Their simplest method is that used by one of them, the scrub-fowl, which lives in the north-east of the continent. Some individuals dig pits in carefully selected sites on a beach where the sun warms the eggs during the day and the sand retains the heat to maintain their temperature overnight. In one place, the birds carefully deposit their eggs in clefts of black rocks which have the same property. On one or two Pacific islands, the scrub-fowl have discovered places where volcanic heat underground performs a similar service for them. Yet others, living in the rainforests inland, rake up fallen vegetation into mounds four and a half metres high which keep their eggs warm by the heat of decay.

The most complex of their techniques is that used by the mallee fowl in the open scrub country of southern Australia. During the winter, the male digs a hole in the sandy earth about a metre deep and four and a half metres across and fills it with vegetation. In the centre of this, he excavates a bowl thirty centimetres or so deep. This will hold the eggs. When the first showers of spring have thoroughly moistened the pile, he covers the whole structure with sand. The vegetation within, protected from the dry air, begins to decay and the mound starts to warm. The female, so far, has played no part in this work. She has been feeding intensively in the neighbourhood, building up in her body the reserves from which she will produce her eggs. When she is ready to lay, the male clears away some of the sand on the top to expose the rotting vegetation, the female lays a single egg in it, and he covers it over again. The male now carefully monitors the temperature of the mound by prodding his beak into it. At the beginning of the season, when the vegetation within is actively fermenting, it may overheat. Then he will kick away some of the sandy blanket to allow heat to escape. As the weeks pass, fermentation and the heat it produces start to dwindle. But the sun has now become more intense. So in order to prevent it raising the mound's temperature too high, a thicker layer of soil has to be heaped over it to shield its interior from the sun's heating rays. As the height of summer passes, the method must change again. Chilling, not overheating, has become the danger, so the male opens up the mound during the day to make the most of the waning sun, and covers it over again at night to retain what warmth it has.

With these methods, varied so expertly with the changing season, the male mallee fowl manages to keep the temperature of his incubator very close to 34°C for several months. Throughout this time, the female has been laying eggs, one at a time. If there is plenty of food about, she may do so every other day. If times are hard then she may only lay once a fortnight. Each time she does so, the male has to dig down to the buried vegetation and cover it over again. He clearly regards the management of the mound as his own particular responsibility and expertise, for if the female comes to the mound to lay at a time when opening it might cause a dangerous fluctuation in its temperature, he will refuse to do so and drive her away. By the time the season comes to an end, the pair and their incubator may have managed to produce as many as thirty-five chicks.

Eggs begin to develop as soon as they are warmed to, and maintained at, the correct temperature, so each egg laid by the mallee fowl starts developing as soon as the female deposits it in the incubator. Their egg is an unusually large one for their body size and has a generous yolk. As a consequence, the chick when it hatches is very fully developed and able to dig its own way up through the sand. After a couple of hours' rest, it runs off into the bush to find food for itself, and after only twenty-four hours it is able to fly. By the time the female lays the last egg of the season, her first chick will have already emerged and left.

Eggs hatching over a period like this present no problems for the megapodes. Nor does such staggered hatching cause any difficulties for such birds as eagles which build inaccessible nests in the tops of trees where their young can remain in safety until they are able to take to the air themselves. But many ground-living birds would be in great difficulties if their chicks hatched at markedly different times. Their chicks initially are too feeble to fly and they are unable to find food for themselves, their parents lead them away from the nest site to other areas where they can collect food and find hiding places. This would be difficult to do if the chicks hatched at different times and were thus at different stages of development and strength. So a female quail does not begin to incubate her clutch of a dozen or so eggs until it is complete, and that may not be for a fortnight after the first was laid. In this way, all her eggs begin developing at the same time.

However, with such a large clutch it is difficult for her to maintain all her eggs at exactly the same temperature. Those at the side of the nest may not be quite as warm as those in the centre. Furthermore, she has to turn the eggs regularly to prevent the membranes within them from adhering to one another or to the shell. So as the time for hatching approaches, all the eggs may not be equally ready. To put things right, the unhatched chicks begin to signal to one another. If you put a doctor's stethoscope to an egg at this time, you may hear clicks coming from within. The neighbouring eggs can also hear them. If they have not yet reached the clicking stage, the sound stimulates them to speed up their development. That this is what happens can be demonstrated by playing recordings of the clicks to one batch of eggs and so inducing them to hatch well before others of the clutch that have been kept individually and in silence.

All in all, nesting time is a particularly demanding and dangerous period in a bird's life. The eggs, even before they hatch, require great care, reducing the time a parent bird can spend finding food for itself and keeping it sitting on the ground or in the branches of a tree where it is exposed to much more danger than it would be in the air. For most birds, however, these risks and labours are inescapable.

But not for all. The female cuckoo has, notoriously, found a way of avoiding them altogether. She induces others to look after her eggs. The European cuckoo bamboozles warblers, dunnocks and robins. The act is done quickly. The female alights on the rim of a stranger's nest, picks up one of the eggs in her beak, flies off with it and then swallows it. She returns immediately and drops in a replacement of her own. Speed is important to prevent the parasitised birds from catching sight of the intruder and becoming aware of the deception, for then they may desert the nest altogether. The cuckoo has no time to settle low in the nest and her egg often falls a little distance. It does not break even if it lands directly on other eggs because its shell is twice as thick as those of the host bird.

The cuckoo is able to produce an egg at such short notice because she has retained it in her body for up to twenty-four hours after it has been encased in shell and is ready for laying. She is therefore able to expel it just as soon as the chance arises. This brings an additional advantage. Retained within the warmth of her body for this extra period, the chick within the egg has already begun to develop. What is more, the incubation it needs is in any case one or two days less than that needed by its hosts' chicks, so the young cuckoo stands a good chance of appearing before the legitimate chicks do. If it does, it will hunch its back and push out the other eggs.

In most cases the cuckoo's egg looks very like its hosts' eggs. Were it not so, the legitimate parents might detect the substitution and throw it out. Although the cuckoo is a bigger bird than the reed warbler or any other of its regular hosts, its eggs are unusually small in proportion to its size, with the result that there does not appear to be a great discrepancy between them. Not only that, but each cuckoo pigments her eggs so that they roughly match those of her host. So female cuckoos that habitually parasitise pipits lay spotted eggs, while those that choose redstarts in continental Europe lay plain blue ones.

The European cuckoo is the most famous bird to behave in this irresponsible fashion, but it is by no means the only one.

The cuckoo family is very large, with a hundred and thirty different species distributed world-wide, and of these nearly half do so. And there are brood-parasites – as they are called – in other families too. In South America, cowbirds of several different species parasitise over a hundred different kinds of small perching birds; in Africa, weaverbirds, whydah birds and honey-guides all exploit the loyal parental responses of others. The trick, clearly, is a very profitable one if you can get away with it. And it is not only birds that do it. The habit is even more widespread among insects.

Ammophila, the caterpillar-hunting wasp, has to keep a close guard over her paralysed prey. She may, for example, lay it down for a few seconds while she opens the entrance to her hole. In that brief time, a fly may dart in and lay its own egg on it. This will hatch quickly and eat both Ammophila's own egg and the caterpillar that was provided for it. Cuckoo-bees make their way into the nests of other kinds of bees and leave their eggs there to be tended and reared by others. Some do so secretively, creeping in quietly among the swarm of busy workers, quickly depositing an egg in a waxen cell alongside those belonging to the builders and slipping out again unremarked. One species, Nomada, ingratiates itself with the colony it parasitises by producing a special perfume that the host bees find particularly attractive. Another, Sphecodes, battles its way into a colony, killing any individual that opposes it. In all these instances, the cuckoo-bee grub, like the cuckoo-bird chick, kills the rightful infants and eats the food-store intended for its victim. In the most extreme cases, such as in the North American Polistes wasp, an invasive female of a different species ejects the dominant queen, and takes over the whole nest, using her odour to persuade the resident workers to help rear her offspring for the whole season.

So unhatched eggs, whether of insects or birds, crabs or reptiles, are threatened on all sides. The most extreme way of protecting them is for the mother to carry them around within her own body.

The male guppy, a small South American fish, has a pair of fins on his underside modified into a gun-like tube through which he fires small bullets of sperm at the female's genital opening. Those that strike their target are absorbed by the female and fertilise the eggs within her. There they also hatch and develop, sustained by the yolk with which she has endowed them. As they grow, they swell and darken, giving her a black triangle on her abdomen just in front of her anal fins. Eventually she releases them. Each babe emerges folded in two, head to tail, the size of a grain of rice, and swims away to hide in the leaves of the water plants.

Among the pipe-fish and their close relatives the sea-horses, it is, surprisingly, the male who becomes pregnant. When they mate, the pair intertwine their bodies. The female extrudes eggs which stick to the underside of the male. The skin beneath them gradually swells until eventually the eggs are seated in little sockets and he carries them until they hatch. This hardly counts as internal fertilisation, but the sea-horse, a shorter relative of the pipe-fish that swims with its body held vertically and not horizontally, has taken the principle of male-brooding considerably further. Three or four days before mating, the male develops a pouch on his belly. When courtship starts, he brings his belly close to that of the female and the two coil about one another. Within five seconds, the female squirts several thousand eggs into his pouch and the two separate. His sperm ducts empty into this pouch so that the eggs are quickly and efficiently fertilised. The pouch-lining then becomes soft and spongy and secretes a nourishing fluid which the babies absorb. Two weeks later, he starts a series of shuddering contractions and tiny miniature sea-horses shoot out from his pouch. They may go on appearing for twenty-four hours until eventually he has given birth to a thousand or more babies.

Such fertilisation within the body is unusual among water-living animals. Most of them simply release their eggs and sperm and rely on the surrounding water to bring them together. But animals that mate on land cannot do this. For them, fertilisation within the female is the rule. Retaining fertile eggs within her body long enough for them to hatch there does not require major anatomical or physiological changes, so it is not surprising that all the major groups of land-living animals contain species that produce live babies. The only exception is the birds. As far as we know, all birds through history have laid eggs. The reason why there are no live-bearing birds might seem to be because the extra load of chicks developing in an adult female bird would make flight difficult.

Even insects, which for the most part lay numerous tiny eggs, neatly packaged in tiny sculptured capsules, in one or two instances produce their young alive. The female bot fly holds her eggs in her oviduct until they have hatched into maggots. Then she injects them into the nostrils of an unfortunate sheep where they quickly start feeding on the membranes lining their host's sinuses. The female tsetse fly retains her young for even longer. She nourishes them with a special fluid which exudes from a nipple on the wall of the pouch in which her larva lies. It breathes through a pair of tubes that it projects from her genital opening. It even sheds its coat periodically, and by the time it is born, it is able to pupate immediately.

This technique brings its own limitations. Lavishing so much care on the offspring, cherishing them until they are well advanced in their growth, inevitably reduces the number of young that a female can produce. Not for the tsetse fly the hundreds deposited as eggs by a house fly. The female tsetse can bear only one baby at a time, and in all her six-month life she can produce no more than a dozen. But the success of her strategy is only too evident to anyone who travels in the areas of Africa inhabited by the tsetse.

The greenfly, another live-bearer, has evolved a way of overcoming even this limitation. During the summer, when there are plenty of food plants in leaf, she produces eggs that, without any attention from a male, are already fertile. They are all female and since they too are also reproductively sufficient unto themselves, they can do the same thing as their mother even before they have left her body. So in effect, the female greenfly gives birth to her children and her grandchildren simultaneously and the offspring of a single pregnant female can, within a few hours, smother a rose bush.

The amphibians have also taken to the habit. Most frogs lay eggs coated in jelly and then abandon them, but some rear their eggs in pouches on their back, in their throat sacs or hold them back in their oviducts.

Among snakes and lizards, the distribution of the technique seems almost arbitrary, and for the moment has defeated the ingenuity of scientists in searching for a genetic or an ecological explanation. Boa constrictors bear live young but pythons don't; rattlesnakes and chameleons do, but cobras and iguanas don't. Typically, the reptilian egg is laid on land and is encased in a shell to prevent its liquids from evaporating, but those eggs that are kept within the female until they hatch have either very thin shells or no shells at all. All reptile eggs contain a considerable amount of yolk, but many of these live-bearing reptiles have developed a way of supplementing it. The eggs become loosely attached to the wall of the female's uterus and blood vessels develop on either side of the contact so that the young are able to absorb sustenance from their mother's blood. In this way, the mother provides her baby with more food over the period of its development than she would be able to pack into a single egg, and what is more, she can do this over a long period instead of being compelled to produce it all at one moment. Once the habit of live-birth has appeared in a reptilian lineage, it tends to lead to social behaviour, as parents and juveniles interact and live together, indicating the evolutionary significance of this mode of reproduction.

The mammals have their own special and characteristic way of fuelling their developing young. The Latin word for 'breast' is *mamma* and mammals produce milk. Those that evolved in the southern hemisphere, the marsupials, start to do this when their babies are very young indeed. Bringing them and the milk together, however, is something of a problem. Antechinus is a tiny Australian marsupial the size and shape of a mouse. The young emerge from their mother's birth canal only a month after conception. They are no bigger than grains of rice. They have strong fore-legs, though their back legs are hardly developed at all and are little more than buds. These tiny pink wet morsels of flesh wriggle their way forward along the underside of their mother's belly through her bristly hair towards a group of a dozen nipples. These are partly shielded by small flaps of skin on each side which form a corridor and may help to guide the babies towards them. The journey is not a long one, a few centimetres only, and the babies quickly find a teat. As soon as one does so, its lips close around it, giving it a firm hold. There the litter will remain, imbibing milk and growing rapidly for five weeks. Towards the end of this time, the mother looks as though she has a bunch of pink grapes on her underside and clambering about the branches or running over uneven ground not only becomes awkward for her but looks distinctly uncomfortable for her young.

But other marsupials live in a way that would make such an exposed method of transport positively lethal. The bandicoots and the wombats are both diggers, excavating long burrows in search of their food. Unshielded babies clinging to teats would soon be brushed off. The females, however, have their nipples enclosed not by flaps of skin but within a deep pouch. What is more, the mouth of the pouch faces backwards so that when the female digs with her forelegs, the earth flies safely past her young as they steadily imbibe their milk within her pouch. The yapok, a South American marsupial, is a swimmer. That way of life might also seem to threaten the lives of her young. But her pouch has a muscle around its mouth which contracts like a draw-string and shuts so tightly when she goes into the water that her young are in no danger of drowning.

The biggest and most famous of all living marsupials, the wallabies and kangaroo, are of course leapers. A young kangaroo would stand little chance of staying on board its mother as she bounded over the land if all it had to cling to was a nipple. The kangaroo's pouch, however, is a deep bag that opens just beneath her chest. A baby – the joey – within it is in no danger of being thrown out no matter how energetically she bounces. But this arrangement means that the joey, when it first emerges from her genital opening has to make a marathon journey. From genital pore to the lip of the pouch may be nearly twenty centimetres. She does nothing to help it. Its forelimbs have tiny claws on them which assist in grasping its mother's hairs and it moves determinedly forward with a movement rather like a swimmer's crawl, turning its head from one side to another with alternate strokes. Since neither its eyes nor its ears are sufficiently developed to be functional at this stage, it probably finds its way to the pouch by smell.

Once it has found a nipple, the tip swells in its mouth so that the young is not able to let go even if it wanted to. After a month or so, its head has developed sufficiently to allow it to open its jaws. Now it can detach itself and move about within the mother's pouch, but it will remain dependent upon her milk until it is eighteen months old. As it grows, the composition of its mother's milk changes to meet the changing needs of its developing body. After about nine months, the baby starts to clamber out of the pouch and hop about at its mother's side, though still returning when danger threatens or when it wants a drink of milk. A month later the youngster has left the pouch for good, but it still suckles, poking its head into the pouch and grabbing a teat. This is the more remarkable since by this time its mother may have already given birth to another tiny baby that has made its way to the pouch and is fastened on to a teat imbibing milk of a quite different composition.

The mammals of Africa, Asia and the northern hemisphere have a quite different way of going about things. Milk is still a vital food for their babies, but the young do not emerge from their mother's womb to take it until they are much more developed. The mammalian egg contains no yolk whatever, but even so, the females are able to nourish their young very effectively within the womb. They use a system similar to that developed by some of the live-bearing reptiles but one that is enormously more efficient. The embryo grows a pad, the placenta, which becomes attached to the wall of the womb. This absorbs nutriment from the mother's blood and conveys it by way of a tube, the umbilical cord, into the body of the baby. With such an effective way of victualling her young, the mother is able to retain them within her until they are so big that their sheer size makes them a burden and the mechanics of getting them out of her body becomes a real problem. Mammalian babies, when they do finally emerge into the outside world, still need their mother's milk to complete the building of their highly complex bodies and they may continue to suckle for years.

Even though mammals lavish such care and attention on their young, they still have the same two options open to insects and fish, crabs and reptiles – whether to concentrate their energies on producing as many young as possible and then let them fend for themselves, or whether to restrict the numbers to very few but look after them carefully.

The only marsupial found in North America, the Virginian opossum, practises the first strategy. She produces as many as twenty-two babies in a single batch. As soon as they emerge, they face their first hazard, the race through their mother's fur to her pouch. She has only thirteen nipples. The first thirteen babies to claim them have won the first of their competitions. All the rest will die. At first her babies are held under her body in a large, gaping pouch. When they get older, they hang on her back like a giant, living fur coat.

The largest litter produced by any placental mammal is even more than the opossum's – thirty-two. That feat is performed by a rabbit-sized relative of the shrews, the Madagascan tenrec. But if the measure of fertility is the number of young that are produced in a season, then the record is held by the tiny meadow mouse (Microtus) from North America. Not only may she give birth to as many as nine babies at a time, but she may have as many as seventeen litters in a breeding season and she is thus capable of producing a hundred and fifty young a year.

Microtus makes her nest underground in a burrow, so she can safely bring her young into the world when they are still at an early stage of their development, and then within a few hours, start on the business of producing the next litter. Her babies' eyes are not yet open, their ears not yet sensitive. They are naked of hair and unable to sustain their own necessary body warmth. But their mother is free of their weight and bulk and she can run into the world outside the burrow to gather food and so continue producing the milk that they need. Many other burrow-livers such as rats and rabbits produce their young at a similar early stage. So do strong aggressive animals which are well able to protect their babies even if they do not deposit them in a den deep underground.

It is those mammals that live more persecuted harried lives that have taken the possibilities of placental development to its ultimate. Surface-dwelling rodents like guinea pigs and agoutis, in contrast to their burrowing relatives, can run almost as soon as they are born, and the calves of wildebeest, born while the herd is on migration, can stand up and trot after their mother within five minutes of dropping to the ground.

So millions of lives are launched in a multitude of different ways – hopping from a pouch, tumbling from a womb, clambering out of a capsule, hammering open an eggshell or falling into the sea, scattered wholesale like seeds. Some will have vastly better chances of survival than others, but for all, the next few months will be the most hazardous of their lives.

TWO

Growing Up

Childhood is an urgent time. For parents, it is only too long a period of encumbrance and handicap during which they are saddled with young who demand continual feeding, who reduce their freedom and regularly expose them to danger. For children, it is too short a time during which they have to build their bodies to the strength and size of an adult and learn all the skills they will need to survive without help.

Few infant mammals grow as quickly as an elephant seal pup. Within minutes of slipping out of its mother's body and freeing itself from its glistening birth membranes, it finds its mother's nipple and is drinking her milk. And what milk! It is twelve times as rich in fats and four times as rich in proteins as the best Jersey cow's milk. The elephant seal mother produces this liquid by transforming the blanket of blubber that lies beneath her skin. And the baby converts it straight back to blubber. There is no time now to spend on the lengthy process of building flesh and bone. That can be done later.

Its mother is in such a hurry because she is out of her element. She has been compelled by her ancestry to leave the sea and come to these beaches to breed, for seals are descended from land-living mammals and although their limbs have become efficient paddles and their bodies are streamlined for swimming, unlike whales and dolphins they have not yet evolved techniques that enable them to give birth at sea. They can only do so out of water. But while she is on the beach with her pup, she cannot feed. The sooner she gets back to the water, the better.

The pup guzzles her milk so steadily and so hungrily that it swells almost visibly. At birth it weighed about forty kilograms. Within a week it puts on another 9 kilos. It lies beside its mother occasionally disengaging from the teat with creamy milk dribbling from the side of its mouth, while she takes a rest and perhaps shifts on the sand to offer her other teat. Neither of them, for this brief period, have anything else to do except to transfer fuel from one to the other.

The beach where they lie is a crowded place. It may be on one of the islands that surround the Antarctic continent or perhaps in Patagonia on the South American mainland. At the beginning of the season, it was claimed by a huge male. He is the biggest of all seals, growing up to four and a half metres long and a weight of two and a half tonnes. The pup's mother was attracted to the beach by his presence. So were as many as a hundred or so other females and now he lies surrounded by them, on guard and ready to do battle with any other male that tries to raid his harem.

These battles are a source of danger to the pup. When the beach-master is angered by a serious challenge, he thunders across the sand, humping and heaving his huge body with surprising speed and taking no notice whatever of what lies in his way. Pups get crushed and killed. Even if a babe and its mother manage to shuffle out of his way in time, the two may be separated and then the pup, while trying to find its way back, may be attacked and badly bitten by other females, irritated at being disturbed. As many as one in ten of the babies born on this beach will die here.

After three weeks, the youngster has tripled or even quadrupled its weight. But now, suddenly, its food supply is cut off. Its mother has given it all the time she can spare. She has lost most of her blubber and she is starving. She must get back to the sea to feed. Laboriously, she heaves her way across the sand, down the beach, into the breakers and swims away.

The pup must now look after itself. If it is extremely fortunate, it may become a double-suckler by finding a female whose pup has just died and who is still producing milk. In that case it will continue to grow until it becomes a gross and bloated two hundred kilos or more.

Having gained weight so rapidly, the abandoned pup then begins to lose it as it uses its blubber to develop the organs of its body. That takes time. The black furry coat with which it was born and which helped it to keep warm during its first days is now sloughed off, revealing the shorter shiny coat which is more suitable for a swimmer. Its globular shape slowly elongates. It stays on the beach for a further six to eight weeks. It may find a few crabs or other invertebrates on the shore to nibble, but otherwise it eats nothing. Nonetheless, it slowly continues to get stronger. Then at last it hummocks its way closer to the sea. The waves lap around it, the deepening water lifts its body from the sand, and it is away. Its brief childhood is over.

The crowded conditions into which the elephant seal pups are born are, if anything, a disadvantage to them, but other young animals rely on great numbers for their safety. Gulls nest together in tightly packed colonies but not because there is a shortage of suitable sites. That is clear because they often do so when there are areas nearby that are equally suitable which they leave entirely vacant. They do so because they and their chicks are much safer in a crowd.

Chicks, squatting on the ground in the open, are in constant danger of attack both from the ground and the air. The colonies of black-headed gulls on the coasts of northern England are raided by foxes. Ten thousand birds and an equal number of chicks constitute an enormous quantity of meat. But the size of the local fox population is governed by the amount of food available, not in times of plenty such as this, but in times of scarcity when the nesting gulls have gone. So there are only a few foxes in the neighbourhood of the colony and they can only raid the margins of it. Chicks in the centre are virtually safe from them. Even those near the edge stand a better chance of survival than if they had been in a solitary nest far away in a place where local foxes can consume everything that is available.

Raids also come from the air. Marauding birds such as herring gulls will grab and swallow a chick if they get a chance. An adult, by itself, will be hard-pressed to repel a determined attack on its young, but in a massed colony, outraged parents join together and surround an intruder in a cloud, shrieking angrily, diving on it and harrying it in a continuous attack. The herring gull has only a small chance of grabbing a nestling in the face of such opposition, and no chance whatever of sneaking up on an unguarded chick unnoticed by adults.

Even with these defences, however, the losses in a colony from a variety of different causes are immense. At one of the major colonies of black-headed gulls in northern England, researchers found that it was a good year if fifteen per cent of the eggs that were laid produced fledged young. But even that was a better success rate than among the nests that were built away from the colony. In one year over three hundred of such solitary nests were monitored. Not one of them produced fledged young.

Gulls travel for kilometres all over the countryside and the coast collecting their food. They fill their crops and then fly back long distances to feed their young. But other animals, particularly those that produce many babies at a time, find it more practical to take their young with them when they forage.

Baby shrews are born in litters of half a dozen or so. They are extremely active little animals and after a couple of weeks, as soon as their fur is grown and their eyes are open, they start to supplement their mother's milk with insects and other invertebrates that they find for themselves. Their mother has a major task in keeping track of them all. If the nest is disturbed and she decides the family should flee, the youngsters behave in an extremely disciplined way. One will seize the fur at the base of its mother's tail and grip it firmly in its jaws. No sooner has it done so than another baby joins on behind the first in the same way and within a few seconds the entire litter has formed a caravan behind their parent. Fast though they run, they all keep in step, so that the whole group looks more like a snake gliding through the undergrowth than a family of young mammals on an outing. So resolute are they at keeping hold of one another that even if you pick up the mother, her babies will still hang on to her in a wriggling furry rope. Once they start foraging, of course, they scatter. But at the slightest hint of danger they join up instantly and dash away like an immaculately trained line of dancers making their exit.

Ducklings at a similarly early stage in their lives also follow their mother as she leads them away on food-finding outings. Indeed, they have a psychological mechanism in their brains which impels them to follow the first large moving thing they see after they emerge from their shells, even if it is not their mother. This reaction was first observed and understood by the great Austrian naturalist, Konrad Lorenz, in the 1930s. He called it 'imprinting'.

Among mallard young, the period during which this process takes place is both precise and limited. It is between thirteen and sixteen hours. If during this time the ducklings see little but pairs of green rubber boots moving around them, then it will be green rubber boots that thereafter they will try to follow, as many people who have raised waterfowl will testify.

This imprinted beacon is not always a visual one. Wood ducks nest in holes in trees. There in the darkness, the chicks can see very little of their parents, but they can hear them and they will follow the sound that they first hear during that crucial imprinting period for the rest of their duckling days.

Nor are ducks the only birds to imprint on their parents in this way. Lorenz did his pioneering work on greylag geese, and rails, coots and domestic chickens also respond in the same fashion. In fact it is likely that the process operates in all young birds that leave the nest very early in their lives and need to follow their parents.

Ostrich chicks are also among them. The male ostrich is polygamous. He has a senior wife who lays the first egg in the scrape he makes in the ground for her. She will deposit as many as a dozen eggs there. But he also mates with several other young females who take turns in coming to the nest and adding their own contributions to the clutch that he is incubating. As a consequence, he may find himself with as many as forty eggs in his charge. It is just as well for him, therefore, that all the chicks, when they hatch, have firmly planted in their minds the notion that they must follow those two massive two-toed feet that are the first large moving things they see.

But the infancy of these chicks is to be more complicated than most. From the beginning, they are beset by dangers. Formidable though the male is and obedient though his followers may be, he has difficulty in defending several dozen youngsters. Hawks may swoop down from the sky and carry one off. Jackals may slip in and grab a straggler. Soon only a small fraction of his original flock is running around his feet. As he leads them about, he may well meet another male who also has an attendant troop of young. If he does, the two males often quarrel. The dispute may become so vigorous that one of them is driven away and flees, striding away over the plains at great speed. His flock has no hope of keeping up with him. For a moment there is indecision among them. But then, once again, they catch sight of a pair of tall scaly two-toed legs and the dominant cock suddenly acquires a bigger flock of babies. Whether he deliberately sought them is not clear. It is possible that he was simply protecting the feeding territory for the benefit of his own offspring. It is, nonetheless, advantageous for him to have his flock increased in this way. If a predator does succeed in grabbing a chick, there is less chance that it will be one of his. So the aim of all parents – the propagation of their genes – is furthered.

Whatever the motivation, these amalgamations may happen several times during the months that he cares for the flock.

He may end up with as many as sixty youngsters from several different broods who, because they hatched at slightly different times, are of several different sizes.

Such large assemblages of young in the charge of only a few adults is known by the French word for a nursery – a crèche. It can be a very convenient way of reducing a parent's problems. The female eider duck is abandoned by her drake just as soon as she starts incubating the eggs. She usually builds on the shores of an estuary and there she sits devotedly, seldom leaving the nest. She does not feed at all. By the time the eggs hatch, she is very hungry indeed. As soon as the chicks are mobile, she leads them away from the nest and down to the shore. Unless hers is the first brood of the season, she finds there a large flock of other ducklings supervised by one or two adult females. These usually include non-breeding birds. Her young quickly join this crèche and start dabbling for small crustaceans and tiny molluscs. Now she too has a chance to eat. But there is nothing here for her. Her favourite food is mussels which she can only find in deeper water. After a few days, her hunger becomes so great that she goes away to feed, leaving her young in the charge of others.

The flock her young have joined may number a hundred. Some eider duck crèches have been counted over five hundred strong. The attendant females are known as aunties, though nannies might be a better term, for although these birds are successful breeders, they may be quite unrelated to the majority of their charges. For this behaviour to be selected by natural selection, it suffices that some of those chicks are related to the aunties. If marauding gulls appear, the aunties sound an alarm and the young cluster tightly around them. If the gull has the nerve to dive at them, an auntie may even grab it by the feet and pull it down into the water.

Chicks in crèches need no more from their aunties than protection. They are well able to collect their food for themselves. But mammal babies cannot be so independent. They need milk from an adult and they need it regularly. Although human mothers, in some places and at some times, hand over their babies to others to suckle, the vast majority of other mammals provide their own babies with most, if not all of their milk. Even so, some mammal parents still find it worthwhile to deposit their babies in a crèche.

In the chilly windswept plains of Patagonia, you may find a burrow about the size of a large rabbit hole. Listen, and you may hear scuffles inside. Watch from a few metres away and you will see an elegant rodent, about the size and shape of a hare but with long stilt-like legs, step daintily towards the hole, nose lifted warily. Locally the people call this animal a mara. Zoologists term it the Patagonian cavy for it is related to the guinea pigs or cavies of the Andes. If, as is likely, this one is a female, she will suddenly whistle a call and out of the hole come a dozen youngsters that bustle eagerly about her, groping with their muzzles for her teats. She skips and twists among them, sniffing their hind-quarters, until at last she finds the two she seeks, her own young, and leads them away to the shade of a bush and there lets them suckle.

All this time, her mate has been standing by, watching the proceedings. If another pair arrives while his mate is with their young, he chases them away, charging at them with his neck outstretched. After about an hour, her two babies scamper back to their den, she rejoins her mate and the two walk off to resume grazing, perhaps a considerable distance away.

But it is not long before another pair appears and summon their own young in the same way. If the crèche is a big one, and some contain as many as twenty infants, then one pair of the adults is likely to be in attendance at virtually all times of the day. They keep an eye on the youngsters and, with the experience and wisdom of their years, will whistle warnings if they sense danger so that all the babies can bolt down the communal burrow to safety. But the adults never go down themselves. Initially one of the females dug the den and soon afterwards, beside its entrance, gave birth to her two or three well-developed babies. They, on their own initiative, crawled inside. Thereafter, as many as a dozen other females will have added their own babies to the crèche in the tunnel in the same way.

Although each mother does her best to restrict her milk to her own young when she visits the nursery, the other babies who surround her when she arrives are so importunate that this is not easy. But the mara's problem is simple compared with that faced by a mother bat. She may have to find her young not from a dozen but from a million.

Every year, female free-tailed bats in Mexico leave their mates and fly two thousand kilometres up to the southern United States. They are pregnant and seeking nursery accommodation of a particular kind. If their babies are to thrive, they must have a cave that is warm, humid and with a temperature that varies hardly at all between day and night. And if the parents are to get all the food they need, it must be in an area where there is a great number of moths and other insects flying at night. There are not many caves that match this exacting prescription. In the whole of the American southwest, there are no more than a dozen or so. But there are a great number of bats and the caves can in consequence become very crowded. Several of them contain in excess of five million bats. One, Bracken Cave in Texas, holds over twenty million.

Inside one of these caves, the air is suffocating with the choking stench of ammonia rising from the layer of soggy droppings that carpets the floor. The heat from the mass of bodies packed within keeps the temperature around 38°C. Venture inside and you would be wise to wear a respirator to filter off the stench, and clothing that will protect you from the steady rain of droppings and urine that falls from the ceiling. The easiest time to see the bats' crèches is at night, for the females do not carry the unnecessary load of a baby when they fly out to hunt. Instead they park them in a special nursery area, packed tightly together so that the loss of heat from their bare bodies is minimised. You may not recognise these crèches at first sight. They look like parts of the wet glistening rock wall which are an unexpected shade of pink. Look closer with your lights and you will see that these great patches are naked babies, packed together like small shiny plums, two thousand to the square metre. It seems impossible that any mother would be able to find her own babe from among such a throng. And until recently, no one thought that she did. A female was believed to come back after a night's hunt and give her milk to whichever baby grabbed her teat first. Now by capturing mothers with suckling babies and making genetic tests on them both, researchers have proved that this is not so.

A returning female alights within a metre or two of where she last left her baby. It is unlikely to be in exactly the same position, even if she could find it, for during the night there is a great deal of jostling among the youngsters and any one of them may have moved about forty-five centimetres or so. As she lands, she calls for several seconds and her baby answers. It is difficult to believe that, among the tumult in the cave, either mother or baby would be able to recognise one another's voices, but bats are famous for their skill in disentangling the echoes of their high-frequency squeaks and using them as a way of navigating. In comparison with that ability, the problem now faced by the mother is simple. Individual calls vary greatly. Slowed down to the frequencies that suit our ears you can hear that they differ in volume, length, pitch and frequency and include squeals and yelps, grunts and trills. When mother and child recognise one another's voices, they try to move closer together across the rock. But that is not easy. As she elbows her way through the massed youngsters, they press around her, trying to snatch a drink of milk. Her nipples are in her armpits, so she keeps her wings close to her sides as far as she can and lashes out with her feet and bites those that pester her. Even so, some manage to grab her teat and snatch a drink as she passes. There is such a commotion that, as like as not, she will fail to reach her baby and will have to fly off and start again. When at last the two do meet, she lifts her wing and her baby nuzzles into her armpit to drink. For five minutes it feeds, she shifting it gently from one nipple to the other. The meal over, she takes off again and flies to another part of the cave where there is less pandemonium. There she suspends herself by her feet and takes a well-earned rest, having done her best to ensure that her single babe, in spite of everything, has received its proper milk ration.

If a young animal is one of a litter or a clutch, it cannot rely on such solicitude from its parents. Many birds deliberately favour some of their young in a way that helps to match the number of babies they rear with the abundance or scarcity of food. Owls, like most birds of prey, start incubating their eggs as soon as they are laid with the result that the chicks hatch at different times. There may therefore be a considerable difference in size between the oldest and the youngest. The first to hatch is inevitably stronger and more vigorous than those that hatch later. When its parent arrives at the nest with food, it pushes the rest out of the way and gets fed first. If there is plenty of food in this particular season, then all the chicks will get fed. If not, the youngest and the smallest will go hungry and before long die. Its emaciated body is then promptly eaten by its elders so no meat is wasted. However heartless and unfair this may seem to human eyes, the end result is more likely to achieve the parent owl's purpose of successfully launching a new generation. Feeding all the chicks equally in a poor season might well lead to all of them dying for lack of adequate nourishment. This way, at least one has the best possible chance of survival.

The task of finding enough food for their offspring dominates the lives of most parents during the breeding season. Sometimes and in some places, the labour is so time-consuming that even the most hard-working of parents cannot manage it unaided.

In Florida, scrub-jays tackle it as a family team. They live in oak scrub, harsh country that is poor both in food and in nest sites. It is not only the breeding pair that occupies the territory around a nest. Several young adults who hatched there during the last two breeding seasons may live there as well. They assist their parents in feeding their new younger brothers and sisters and in defending them from predators such as snakes. The majority of these helpers are young males. The young females tend to fly off and look for mates elsewhere. If the group is really successful and energetic, they may be able to expand the family property. Eventually it may become so big that one of the young sons may be able to set up on his own in one corner of it. He will then be in a good position to take over much of the rest of the property when his parents die. But about half of these helpers will never breed. Their lives have been devoted to the welfare of the next generation – not, it is true, to their direct descendants, but to their younger brothers and sisters and therefore potentially to their nephews and nieces, with whom they share genes.

This sort of cooperation within families is much more widespread than was recognised in the past. Moorhens, wrens and woodpeckers all include species that, in some circumstances, behave in this way. Ten to fifteen per cent of all the birds in Australia do so as well. And so do some mammals.

Marmosets, tiny monkeys that live in the canopy of the South American rainforest, have a very hard time raising their babies. They have to be constantly on the move, seeking the fruit and insects on which they live, but their babies, usually twins, are particularly large and have to be carried piggy-back until they are quite old. Their mother inevitably expends a great deal of her energy in providing them with milk and the task of carrying them as well is too much for her. So for a lot of the time, their father takes on the job. But even he needs help, for it is not easy to grab an insect or pluck a dangling fruit if you have a couple of half-grown babies on your back. So some of the young of the pair stay with their parents for several years and take turns in transporting the new infants. Father even allows quite unrelated youngsters to join the family group if they will take on some of the work of baby-carrying. As a result, there may be as many as nine adults in one of these family parties. But only one female and male among them breed. The male does sometimes mate with one of the young female helpers but this, for reasons we do not understand, never seems to result in pregnancy.

Elephants also collaborate in looking after their young. All the adults in a herd are female. Their leader is the oldest and wisest of them. The rest are her sisters, daughters and grand-daughters. The bulls lead more or less solitary lives outside the herd. The birth of a calf is a great event in this community. The females, young and old, crowd around the newcomer, rumbling murmurs among themselves, caressing it with their trunks and helping to free it from its birth membranes. The baby is able to walk after an hour and to keep up with the herd when it is on the move, but it is nonetheless very unsteady on its feet and needs continual help. Climbing up a steep bank or trying to extricate itself from a mud hole is likely to produce squeals of distress from the infant and the adults nearby will rush over to see what is wrong. As time goes by, the mother seems to become a little blasé about these shrieks for attention and lets the still enthusiastic young females fuss over the infant.

The baby continues taking its mother's milk until it is at least two years old. Just as a human baby may draw comfort from sucking a dummy, so occasionally a baby elephant will sidle up to a young female and suck her milkless nipple. The adult seems to enjoy the experience just as much as the baby does. Should a youngster become orphaned at this early stage in its life, one of its aunts, if she is in milk herself, may allow it to suckle alongside her own baby and, in effect, adopt it.

Like so many baby mammals, young elephants spend a great deal of their time playing. They engage in butting matches; they chase one another around the great moving pillars of their aunts' legs; they wallow ecstatically in mud. Watch them doing so and you can have no doubt that they are enjoying themselves hugely, just as human children do in a playground. But play has a serious and valuable purpose. It is a way of learning. One of the first things an infant elephant has to discover is how to use its trunk. When it is only a month or two old, the long dangling object on the front of its face is obviously a puzzle to it. It will shake its head and observe how this curious appendage flops about. Sometimes it trips up over it. And when it goes down to a water hole to drink it crouches down and awkwardly sips with its mouth. Not until it is four or five months old does it discover the remarkable fact that water can be sniffed up into a trunk and then, if you blow out, you can hose it into your mouth. And that discovery, of course, leads to a whole new set of possibilities for games.

Lion cubs too play games that help them master the skills that will be essential for their success in later life. As their mother lies dozing, one of them will suddenly pounce on the black tassel at the end of her twitching tail, using the same kind of actions it will need to pounce on a small prey animal in years to come. They fight with one another, too. Even when they are only a few months old, they have claws and teeth quite long and sharp enough to damage one another. But before they start on their game, they signal that this is not to be a serious quarrel by walking in a stilted exaggerated fashion. Then when they strike out at one another, they keep their claws sheathed.

As they get bigger, so the lessons become more realistic and more serious. A lioness, having caught a gazelle, may not kill it but drag it back alive to her cubs and give it to them so that, crippled though it is, the cubs may have a little practice in how to bring it down. A mother otter will bring a half-dead fish and give it to her young to play with in a pool so that they may practise the dives and swoops that are needed to be a successful underwater hunter.

As infants grow, their bodies not only enlarge but change both in shape and colour. Many will have been wearing a special costume for their childhood which has given them particular concealment. Fawns have dappled coats that match the broken light of a woodland floor. The downy chicks of terns and gulls are so patterned that they are almost invisible crouching on shingle. Both adult birds and their young stake everything on the effectiveness of this camouflage. When an intruder approaches, the parents fly off and the chicks sit tight no matter how close it comes.

The young of the European wild pig, unlike their parents, are striped, camouflaging them in the dappled shade of the forests where they naturally live. They can behave in a different way to young birds. Sometimes, it seems that they have no confidence in their invisibility. If they are disturbed, they and their parents may bolt together. Their juvenile costume, therefore, may have a dual purpose. Perhaps it is also a distinctive signal which ensures that the parents do not eat the piglets but treat them with the restraint and solicitude due to babies.

The young of the whiteface butterfly fish that live on coral reefs seem to use this system too. Initially, they have ultramarine flanks patterned with white concentric lines – totally different from their parents which are spectacularly dressed with alternate parallel stripes of yellow and blue. In this way the youngsters appear to make it clear that they are not yet mature enough to be considered rivals for territories or breeding partners and are allowed to feed on the reef beside their parents for the several months that it takes them to grow to maturity.

So childhood approaches its end. For some, such as elephants and lions, the process is a gradual one, as the youngsters drift farther and farther from their parents and become less reliant on them for food. For others, the transition to independence is only too sharp. Young albatrosses on the Leeward Islands spend many days flapping their wings to exercise their muscles and develop their strength, but when they launch themselves into the air, they have to get it right first time. Many do so, and with their aeronautical skills improving visibly with every wingbeat, they set off across the sea, climbing steadily.

Others are not so successful. Flailing their wings inexpertly, they drop to the surface of the sea. There, tiger sharks have gathered, as they do each year awaiting this sudden abundance of food, and they rise with their jaws agape to seize them. Some of the youngsters are engulfed in a single bite. Others, struggling to get air-borne again, are driven backwards by the bow-wave of the surging sharks as they break the surface. They peck valiantly at the monsters' pointed snouts, frantically paddling with their feet and beating their wings. One rises. The shark catches it by its feet, but as it tries to get a better grip the young bird is released and flaps into the air over the shark's back. Before the shark can turn to make a second attack, the young albatross manages to make just enough height for its feet to clear the water. It has survived the first crisis of its independent life.

The young free-tailed bats must also eventually leave their nursery. While they were there, inside the cave, they were comparatively safe. There, less than one in a hundred died. But when they take their first flight into the outside world, their losses begin. In the sky above, bat hawks circle, hover and pounce. On a tree-stump beside one cave-mouth, a raccoon sits swatting them down with its paw, munching the little bodies and discarding the skinny wings onto a growing pile on the ground beside it. As the sun sinks, the young bats stream from the cave-mouth like smoke and set off on the first stage of their long journey south.

Ten million bats are born each summer in Bracken Cave, Texas. Before a year has passed, seven million will be dead. Such are the perils of childhood.

THREE

Finding Food

Animals have to kill to feed. They cannot, like plants, build their bodies from nothing more than minerals drawn from the earth and gases extracted from the air. They must eat plants. Some consume them directly; others do so indirectly by eating the bodies of plant-eating animals. Neither plants nor animals welcome being eaten. So finding food, for an animal, can be a demanding and continuing trial.

Some plants, most notably grasses, can be consumed without much difficulty and with few penalties, but a surprising number defend themselves. That becomes only too clear if you find yourself hungry in a tropical rainforest. You are surrounded by the most varied and abundant assemblage of plants in the world, so there, if anywhere, it should be easy enough to collect a vegetarian meal. But the trunks and stems around you are armed with savage spines and hooks. Roots are laden with poison. Leaves are studded with stings. You realise then, well enough, that making a meal of plants can demand both skill and knowledge.

Plant-eating animals, of course, have both. Woolly monkeys live very largely on leaves and they spend hours every day sitting high in the canopy, thirty metres or more above the ground, plucking leaves and cramming them into their mouths. But they do not do this carelessly. They inspect each leaf, turning it over, sometimes smelling it, rejecting this one, selecting that. They do this because most of the forest trees protect themselves against molesters with a poisonous sap. The poison develops a little time after the leaf has sprouted, so the monkeys can avoid the worst of it if they eat only the youngest leaves. But even then, they cannot avoid it altogether and after a while their digestions can no longer tolerate it. So they move off that tree and settle on another of a different kind. Its leaves too will have their own toxin, but since it will be chemically a little different, the monkeys can take another course of their meal there.

Some plants have more virulent poisons. The North American milkweed, if it is damaged, exudes a milky sap. As it oozes from the wound, it solidifies and so repairs the injury. It also protects the plant in a more general way for it has such a bitter taste and is so poisonous that most animals will not eat the plant. Cows and deer and horses leave it well alone. But some insects have found a way of eating the leaves. Beetles, landing on one, immediately cut through the mid-rib. The latex, flowing to the site of the wound, drips harmlessly to the ground and the beetle then munches the leaf tissues beyond the cut which the latex cannot reach. Some species of caterpillar not only sever the veins in this way, they also gouge out a circular trench on the underside of the leaf and then feed only within this protective moat.

The caterpillars of the monarch butterfly, surprisingly, are able to feed on milkweed without taking any of these precautions. Together with a very few other insects, they have developed an immunity to the poison. This remarkable biochemical achievement brings them considerable rewards. Since most other animals avoid milkweed, the monarch caterpillars usually have the leaves all to themselves. They also store the poison in their tissues and so themselves inherit its protection, for birds find the caterpillars just as distasteful as grazing animals found the milkweed. Since the caterpillars have very bright distinctive colours, the birds quickly recognise that they are inedible and leave them alone. The molecules of poison remain unchanged and potent, even after the caterpillars have reconstructed their bodies and turned into butterflies. Since they still have the milkweed's toxin within them and, like the caterpillars, are vividly coloured, they too are left alone by birds. In some butterfly species that feed on toxic plants, the female is able to detect from a male's odour how much poison he has sequestered in his body, and decide whether or not to mate with him. If she does, the plant poison he transfers to her with his sperm will help protect her and her offspring against being eaten by spiders and other predators.

But things are not always made so difficult for vegetarians. Sometimes plants actually encourage animals to feed on them. They need to have genetic material, pollen, transported from one individual plant to another and they are prepared to sacrifice a substantial proportion of it as food to those that will do the job. To advertise the fact, they surround the pollen and the anthers that produce it with the vivid petals of a flower.

The bumble bee has developed complex machinery for collecting pollen. The hairs on its furry body are covered with microscopic hooks which pick up the slightly sticky pollen grains as the bee busies around and within the flower. When it flies off, it starts to tidy itself up, sweeping its fur clean with a bristly comb that fringes the lower half of each hind leg. Having done that, it reaches one leg across to its equivalent on the other side, removes the accumulated pollen from the comb with a stiff brush that sprouts from the end of the leg and transfers it into a deep bowl lying on the outer surface of the opposite thigh. This is hedged around by a palisade of long bristles. The bee then kneads the pollen into these baskets with its middle legs, moulding it around the slender peg that projects from each of them, so that when it flies back to its nest after a successful trip, it has a brilliant yellow button of food attached to each thigh.

Anthers, in most flowers, release their pollen by splitting lengthways and allowing it to shower out. Those of the nightshade, tomato and several other plants, however, do so through a small opening at the tip, through which the pollen falls a few grains at a time. This is far too sparing for a bumble bee. When one lands on such a flower, it seizes the clumped anthers with its six legs and then vibrates its muscles so vigorously that its whole body shakes with a loud buzzing noise and pollen pours out of the anthers like salt from a vigorously-wielded salt cellar.

Some of this pollen is inevitably brushed off when a bee visits another flower, but bearing in mind that only a few grains are needed to bring about cross-fertilisation, and that a bee on a single journey may collect two million, it is clear that the price a plant pays for this transport is a very extravagant one.

Pollen grains, which consist largely of precious genetic material, are quite demanding for a plant to produce and many flowers offer, either in addition or as an alternative, a payment that is considerably cheaper for it is nothing more than sweetened water, nectar. They manufacture this liquid in nectaries that are usually located in the farthest depths of a flower. There the nectar does not evaporate as quickly as it would if it were in a more exposed position. Neither does it lose its sweetness by becoming diluted with rain water. Such a placing also ensures that insects coming to feed on it, even if they seek nothing else, are dusted with the pollen that the plants require them to carry. Bees collect nectar as well as pollen, sucking it up with their long tubular mouthparts and carrying it back within their stomachs to their nest, there to store it in their combs as honey.

The limitation of this sort of food, from an animal's point of view, is that it is only available during the short season when a plant is in flower. So nectar-feeding butterflies in their adult form can only be active during the summer and bees must work frantically hard, gathering what pollen and nectar they can while it is available, and storing what they do not need immediately, to nourish their colony through the barren season.

The honeypot ants of central Australia have a similar storage problem. They solve it by using some of their numbers as jars. These specialised workers, known to entomologists as repletes, never leave the nest but inhabit galleries two metres down in the red earth. When workers which have been out foraging return with crops full of honeydew and nectar, they feed it to a replete, which swells until its abdomen, once no bigger than a grain of sand, has swollen to the size of a large pea. Its tiny head and legs project from one side but it can no longer crawl about. All it can do is to cling to the roof of its gallery and there hundreds of them hang in rows. When the dry season comes and food is scarce, the active workers of the colony visit this living larder and caress the repletes with their antennae until they regurgitate droplets for them.

Most animals living in lands where there is a flowerless season and lacking a storage system can only treat pollen and nectar as summer supplements to their menus. In South Africa, little rock mice collect nectar from – and pollinate – some species of Protea which obligingly bear their flowers close to the ground and facing downwards. In Madagascar, geckos lap nectar from palm flowers. In Europe, blue tits collect it from the crown imperial fritillary, the only European plant known to be pollinated by a bird. Some species of fruit bat also sip nectar when it is available. The plants that rely on them to do so only open their flowers at night. They are pale in colour so that they are more easily seen in the darkness.

Because animals such as these feed on other things at other times of the year, they cannot develop specialised nectar-collecting apparatus, for that would make feeding on other substances difficult if not impossible. In the tropics, however, flowers of one kind or another can be found throughout the year, so the animals are able to make pollen and nectar the mainstay of their diet and have evolved highly efficient organs to collect it.

Several groups of birds have done this. The lories, a branch of the parrot family, have acquired a tongue with little papillae on its surface that can be erected to form a brush with which to sweep up nectar. The hummingbirds in South America, and the sunbirds in Africa, are equipped in a different way. They do not have a brush tongue but an extremely long one divided into two from its middle down to its tip. It was once thought that the birds used it as though sucking up a drink through a straw. In fact they lick up the nectar by flicking their tongues rapidly into the flower, in the case of hummingbirds as swiftly as thirteen times a second.

It is not in a plant's interest to provide copious and unlimited supplies of nectar. Its purpose is better served if a pollinator, instead of making one visit and drinking deep, calls repeatedly every hour or so, day after day. It will thus be dusted with successive batches of pollen as they mature and will deliver them to other flowers on different individual plants elsewhere in its territory. Heliconia, a kind of wild South American banana, produces long hanging stems with lines of triangular spiky flowers on each side. The flowers mature in sequence, starting with the oldest at the top. Each produces only a few drops of nectar at a time. The hummingbird which feeds on it must therefore visit many plants, one after another. Once a nectary has been drained, it takes a little time to refill. If the hummingbird comes back too soon, there will not be enough food to compensate for the energy it had to expend in getting there. On the other hand, if it delays too long, a rival bird may have collected the prize. So a hummingbird that specialises in feeding from Heliconia has to patrol a whole group of plants, visiting each hanging spike of blossoms in strict rotation on a carefully timed schedule.

Nor is doling out rewards in tiny instalments the only restriction that a plant may impose in order to compel its pollen-carriers to provide the service it requires. Pollen delivered to a plant of a different species is pollen wasted. Far better that its messengers take it only to plants of exactly the same kind where the genes it carries will unite with eggs and form seeds. So flowers keep their nectar behind locks to which only a small group or even a single species has the key, which is of such a specialised design that its owner finds it difficult if not impossible to use on any other flower.

Some South American flowers are shaped like curved trumpets. Only hummingbirds with curved bills can drink from them and their beaks fit into them as accurately as a curved dagger sliding into its scabbard. A Madagascan orchid, Angraecum, secretes its nectar in a green tubular spur thirty centimetres long that hangs from the lip of a star-shaped snow-white flower. Only the bottom centimetre of this contains nectar. When in the nineteenth century naturalists first examined it, they were mystified as to how any animal could feed on it. Charles Darwin, when he was shown the flower by baffled botanists, confidently predicted that, bearing in mind the flower's colour and size, a night-flying moth would be its pollinator and that, unlikely though this might sound, it would have a proboscis thirty centimetres long. Entomologists at the time said that such an idea was ridiculous. Forty years later, just such a hawk-moth was discovered and to commemorate Darwin's prescience, it was given as part of its scientific name the words *forma praedicta*.

But such obstacles are not sufficient to defeat every hungry animal. Where there are locks there are lock-pickers. Carpenter bees use their saw-jaws to cut holes in the sides of flowers to get at the nectaries. A whole group of South American birds related to tanagers are known as flower-piercers, for they do exactly that. The upper half of their beak has a hooked tip with which they latch on to a flower. They then use the shorter needle-sharp lower mandible to cut a slit into the flower through which they insert their tongue. This, while nowhere near as long as a hummingbird's tongue, is quite long enough from this position to burgle the nectaries.

Later in the season, plants offer animals food of a different kind and for a different reason. After their flowers are fertilised, their seeds develop. These too need to be distributed, for it is better that the next generation should grow away from the parent plant, where they will not suffer from its shade or be deprived of sustenance by its roots. Some seeds are light enough to be blown away but bigger seeds can only be shifted by animals and once again plants produce special food as payment for the job.

The fig tree embeds its multitudes of seeds in globes of sweet pulp. When one comes into fruit in the rainforests of South America, animals from kilometres away flock to it. The figs themselves are hardly bigger than acorns, but there are huge quantities of them and all kinds of animals find them irresistible. An atmosphere of carnival surrounds such a tree. Irritations that occur between animals at other times are forgotten in the general feasting. Woolly monkeys and howlers, spider monkeys and capuchins, tamarins and marmosets scramble around one another reaching for the fruit. Parrots and macaws grab them with their beaks. Toucans and hornbills collect them one at a time, throwing them up in the air and deftly catching them at the back of their throats. The seeds with which each fruit is packed pass down the banqueters' digestive tracts unharmed and are excreted, with luck, some way away.

The very digestibility of the fruit, however, may be a disadvantage as far as the plant is concerned. Because fruit is not very nutritious in proportion to its bulk, animals that eat it must consume a lot. If they are not to develop large unwieldy stomachs, they have to digest it quickly and get rid of the waste. And that, indeed, they do. Some fruit-eating birds can swallow fruit and only five minutes later squirt out the remains. Consequently, many of the seeds land close to where the diners collected them and the plants' purpose is largely thwarted. But some at least remain in the animals' stomachs and guts long enough for them to be carried away when the sated diners depart.

Some animals will tackle much bigger seeds. The nutmeg, complete with its edible rind, is almost as big as a hen's egg, with a diameter of about five centimetres. The green imperial pigeon, uniquely among birds, can unhitch its lower beak and expand its mouth not only vertically but horizontally and swallow a nutmeg that is slightly larger than its own head. Seeds of such a size remain in its gizzard for a little time while the rind is stripped off them and then later, perhaps when the bird has returned to one of its habitual perches, the nutmeg is regurgitated and falls to the ground. But the pigeon's adaptation to such a diet of large objects is so good that seeds only a little smaller than a nutmeg will pass right through the bird and be ejected together with a small quantity of droppings which will help them grow on the forest floor.

Seeds, of course, are much richer in nutriment than any fleshy coating. They are packed with concentrated food to fuel the young plants throughout the first stages of their growth until they are able, with their leaves, to manufacture food for themselves. But whereas the plant makes the flesh of its fruit easily available to animals, it goes to great lengths to protect its seeds from pilferers, usually enclosing them in armour of some sort.

Animals, on the other hand, do not allow such a rich source of food to remain untapped. Crows and tits hold small seeds in one foot and hammer them open with their beaks. A hawfinch has such a powerful bill that with straightforward muscle power it can crack a cherry stone or even an olive pit. The Brazil nut is one of the best protected of all nuts, but even that can be dealt with. The agouti, a long-legged rodent that forages nervously on the forest floor, has chisel-sharp front teeth that can cut right through the Brazil nut shell and extract the rich kernel.

Even so, the Brazil nut does not lose the contest entirely. The tree produces its nuts in groups that fit neatly together like the segments of an orange and are packaged in a box. When the box falls from the tree and hits the ground, it splits open, scattering the nuts. If the agouti finds one, it is likely to find far more than it can eat at a single sitting. It industriously collects them, stuffing them into its cheek pouches. Then it buries those it cannot eat, one at a time, in different parts of its territory to be retrieved later when times may be harder. But the agouti's memory is fallible. It does not always remember where it has buried each one. So in spite of the tree's failure to make its seeds totally impregnable, it has succeeded, at the cost of a percentage of its seeds, in getting some of them distributed.

Plants have another way, too, of deterring animals from destroying their seeds. They pack them with poison. Strychnine, one of the deadliest of all poisons, comes from the seeds of a tall evergreen tree which grows in tropical Asia. Its fruit are about the size and colour of small oranges, and squirrels and hornbills feed on the fleshy pulp. But both animals take great care not to crack and so destroy any of the disc-shaped seeds.

Macaws, which specialise in eating seeds, have to deal with this problem very frequently. The birds normally live in isolated pairs, but at some times of the year they congregate in great numbers on particular sites on river banks. There they gnaw the soil. They are not digging nest holes. Nor are they indulging in some form of display. They come to these special places to gather specific minerals such as kaolin which neutralise the poison they have absorbed from the seeds they eat at this season of the year.

Just as seeds provide a rich feast to those that have the equipment to deal with them, so do their animal equivalent, eggs. And special tricks and tools are needed to open them too.

The kusimanse, a dwarf mongoose from West Africa, confronted with a chicken's egg, will put its fore-legs over it and with a vigour that would not disgrace an American footballer, hurl it backwards through its splayed hind legs. Sooner or later, this hits something and – sooner or later – it cracks. The Egyptian vulture, finding a clutch of ostrich eggs, picks up sizeable stones in its beak and with a nod of its head, tosses them in the general direction of the nest. What it lacks in aim, it makes up for with persistence and it too usually manages to break open one of the eggs. When it does so, it has a bigger meal than it can manage by itself and many another animal comes to lap up the spilt yolk.

The African egg-eating snake has a most remarkable egg-opening tool. Like most snakes it can disarticulate its lower jaw from its upper when confronted with a particularly large meal, but it does more than that. The vertebrae in its backbone just behind its head have spikes on their lower surface which project downwards into its throat to form a small saw. As the egg is pushed down its gullet by muscular contractions, this saw slits open the shell. The yolk is squeezed out and the crushed shell, usually still held together by its membranous lining, is then neatly regurgitated.

Snails, from a hungry animal's point of view, present much the same problem as eggs, for they too are succulent morsels enclosed in hard shells. But for the thirst snake of Guyana they are a favourite food and it has specially modified jaws with which to deal with them. In particular, it has a lower jaw so loosely connected with the upper that it can be pushed forward like a long narrow spoon. The snake grabs a snail and while holding it firmly with the teeth on its upper jaw, it pushes its lower into the shell opening. Hooked teeth on the tip snag into the snail's body and with a twisting movement the snake neatly pulls it out and swallows it.

Birds also know how to deal with shelled molluscs. The Everglades kite in Florida picks up snails and carries them off to a feeding perch. There, grasping the snail firmly in the talons of one foot, it waits. Eventually, the snail slowly and cautiously puts out its head – and is immediately seized by the kite's beak and, with a quick tug, jerked out of its shell. Thrushes deal with the problem more vigorously. They take garden snails, which have relatively thin shells, and holding them in their beak, smash them on a rock, which birdwatchers call, appropriately, an anvil.

Wading birds collect great quantities of small molluscs from sandbanks and mud flats when the tide retreats. Each can be extracted from its shell with a flick of the head. Knot and dunlin collect them, as well as worms and insect larvae, by probing and ploughing through the mud with their bills and detecting the morsels they seek by touch. This can be a dangerously exposed occupation. Out on the flats there is nothing to hide behind and the birds must keep a sharp look out for trouble. Not surprisingly, therefore, they work in flocks so that there are always some individuals with their heads up, able to sound an alarm the first moment their safety seems uncertain.

Redshank and ringed plovers also feed out on the flats. They, on the other hand, seem to take more risks, for they operate singly. Why should they forego the safety of the flock? The answer concerns their food. Instead of picking out creatures that are buried, they collect spire shells and crustaceans that lie on the surface and they find them by sight. If these animals detect vibrations in the sand or ripples spreading across the shallow pools, they swiftly disappear into the mud where the redshank and the plover cannot see them. If these birds foraged in groups, the actions of one would disturb the prey of another and together they would collect very little. They are forced therefore to work in isolation, even though it is riskier to do so.

Insects hiding on the trunks and branches of trees represent an equally rich harvest for those animals that know how to gather them. Moths and bugs and spiders conceal themselves in crannies of the bark. In British woodlands, the tiny tree-creeper collects them, clinging to the tree trunks with its long slender toes and curved claws, propping itself up with its stiff tail pressed hard against the bark. This stance is best suited to upwards travel and the bird moves in a tight spiral round the tree so that no part of the bark is left unscanned. Having got to the top, it flutters down to the base of another tree and starts all over again. The nuthatch, on the other hand, in spite of being a slightly larger bird, is rather more versatile. It does not rely on its tail and the downward pull of gravity to give it a secure grip. Its toes are strong enough to hold it on the bark, no matter which way it is facing and it hops about in all directions with sure-footed nimbleness.

Both these birds eat only that which can be picked off the surface or tweezered out of the crevices. There are other meals to be had that lie deeper beneath bark or in the wood itself. These require different techniques and different tools. Woodpeckers have a particular taste for beetle grubs. They listen for them, cocking their heads to one side to catch the sound of the grub chewing its way through the wood which, since wood is its food, it does most of the time. Once detected, the bird swiftly chisels a hole and then unreels its enormously long tongue. This can be extended for as much as four times the length of its beak. It is stiffened by a slender bony rod housed in a sheath that runs from the back of the beak, behind the skull and over its top to the front of the face. In some members of the family, it goes even further and coils around the right eye-socket. In the American flickers, close cousins to woodpeckers, the tongue is so long that it extends beyond the eye-socket into the upper beak, entering through the right nostril so that the bird can only breathe through its left. Flickers use this amazingly long tongue to feed on ants. They lubricate it with a saliva that is sticky so that the ants adhere to it, and alkaline so that the formic acid of their stings is neutralised.

There seems to be no better tool for extracting ants and termites from their nest than such a tongue as this, for birds are not the only animals to have developed it. So have mammals. Several groups of them specialise in this diet and each has evolved a long sticky tongue entirely independently: a marsupial, the numbat, in Australia; a distant relative of elephant shrews and tenrecs, the aardvark in Africa; the pangolins of Asia and Africa which are covered in a mail of horny plates so that they resemble giant animated fir-cones; and the three very different ant-eaters of South America, the gazelle-sized giant of the savannahs, the squirrel-sized pygmy from the forest canopy, and the monkey-sized tamandua from its mid-storey.

How long it took the different ancestors of these very different animals to evolve such tongues we do not know for there is no fossil evidence of any antiquity to tell us, but it must have been several million years. One animal, however, has acquired the ability to fish termites out of their nests very much more swiftly. It may have even done so within the last few centuries, though no one can know for sure. The chimpanzee has not waited for evolution to shape its body into a specialised tool. It has used its intelligence and dexterity to make one. It plucks a long grass stem, carefully strips it of any side leaves it may have and then pokes it down one of the entrance holes of a termite nest. The worker and soldier termites attack it, as they will attack any intruder. They clamp their jaws on it and hang on, martyring themselves for the sake of the colony. The chimpanzee then pulls out the grass stem and picks off the termites with its teeth, smacking its lips with pleasure. Fashioning the tool is a complex affair, which young chimpanzees learn entirely by carefully observing the behaviour of their elders. There is no evidence that adult chimpanzees deliberately teach this difficult task to their offspring.

Insects can be gathered not only from their nests and tunnels, but from the air. They, indeed, were the first animals to master flight and they did so over two hundred million years before birds managed to do the same thing. Still today virtually all insects at some stage of their lives take to the air. Termites and ants do it seasonally when the time comes for them to mate, disperse and set up new colonies. And vast numbers of other insect species fly throughout their adult phase. So, naturally, many insect-eating animals pursue them in the air.

During the day, swallows hawk for them at low levels. At higher altitudes, swifts trawl for them with open beaks. A pair of swifts with a family to feed may catch twenty thousand insects in a single day. At night, frog-mouths and nightjars feast on them. Many of these insect-feeding birds have a line of bristles around their beaks which at one time were thought to channel weakly-flying insects into the open beak, but now it seems more likely that they do no more than protect the birds' eyes as they deliberately plunge through clouds of insects.

Bats, which invaded the skies even more recently than the birds, also exploit this immense source of protein. The ten million adult Mexican free-tailed bats in Bracken Cave remove one hundred tonnes of tiny insects from the skies of Texas every night.

But flying insects have much more ancient enemies. Almost as soon as they took to the air, other invertebrates – the spiders – started setting traps for them. All living spiders produce silk. It is a liquid protein that they squeeze from little nozzles at the rear of the abdomen which hardens as it meets the air. Any one spider can produce different kinds of silk from different spinnerets. Some kinds are stronger than steel wire of the same diameter and are the strongest of all known natural fibres. Spiders use it for all kinds of purposes – making egg sacs, lining nests, weaving tents for their babies and as safety lines when they jump. But their most ingenious use is in the making of insect traps.

It is the females who nearly always are the active hunters. The bolas spider spins a single filament which she weights at one end with a drop of glue laced with a scent that resembles that produced by a female moth. When she hunts, she hangs below a twig and whirls it towards an approaching insect. If her cast is a good one, and if the potential prey is attracted by the odour, then the thread entangles the insect and she hauls it in. Another, the scaffold-web spider, rigs a whole series of sticky threads from the branches of a bush down to the ground and hauls each one so tight that if an insect, either walking on the ground or flying a little way above it blunders into one of them, the thread breaks and the victim, stuck to it with glue, is hoisted into the air. There it hangs until the trap-maker hauls it in and consumes it. Yet

65

another, the gladiator spider, spins a small skein of silk and holds it between her four front legs. When an insect comes near, she straightens out her legs and holds the net above her head and so entraps it.

The most complex trap of all, however, is the one that is the most familiar, the orb web. It is usually stretched across a flyway where there is likely to be a frequent passage of insects. To construct one, the spider clambers up one side of the gap and lifts her abdomen into the air. From one of her spinnerets, she extrudes a thin filament of silk. So fine is it that even the slightest movement in the air will catch it and lift it. As it floats away, the spider continues to spin until there may be as much as a metre of thread hanging in the air. Eventually it may be so long that it reaches the other side of the gap and becomes entangled in a leaf or a twig. As soon as that happens, the spider stops spinning, turns round, pulls the filament tight and secures it on her side.

Gingerly, she crawls across this pilot line, trailing behind her a thicker, stronger thread. Once that too has been firmly fixed, she starts to erect lines crossing the gap like spokes of a wheel. On these she fixes a spiral mesh. For this, she uses a thinner silk from a different spinneret. As it emerges, it is given a continuous coat of glue. When a length has been rigged between one radial cable and another, she gives it a twang with one of her legs and the glue breaks up into a line of beads. As they form under the forces of surface tension, they drag the silk into little bundles within them. So the filament is pulled tight. But the thread now has great stretchability. If the wind blows hard or an insect travelling at speed flies into the web, the bundles of thread in the glue globules unwind so that the filament can extend its length up to four times and still does not break. And when the tension is released, it contracts again.

With the spiral in place, the trap is set. The spider may wait for developments sitting in the centre of the web or she may retreat to the side and lurk there with one of her eight legs resting on a cue-line through which she will feel movements in the trap. Winds may blow through the web, causing the closely-set mesh to belly outwards, but the web will recover. If an insect blunders into it the owner will rush out, deliver a swift poisoned bite to her victim and then wrap it in bonds of yet a different silk and carry it away for consumption at leisure.

This complex and beautiful structure seldom lasts for more than one night. Captures may damage it and the beads of glue exposed to the air may lose some of their stickiness. Then the spider will roll it up and, so as not to waste the valuable protein of which it is made, eat it. And the next night, she will spin it all over again.

All these small invertebrates – flying insects trapped in webs by spiders, larvae concealed in bark picked out by woodpeckers, molluscs hidden in mud gathered by wading birds, termites licked up by anteaters – all are harvested with little more than the effort expended by those animals that sip nectar and munch pollen, or gather fruit and chew leaves.

Animals that feed in these ways have a relatively easy time of it. But not all meals are so readily secured. The most nutritious of all foods, meat, comes from bigger backboned animals. So meat-eaters have to be vigorous and athletic hunters. To collect their food, they have to use a completely different set of strategies.

FOUR

Hunting and Escaping

Sea-lion pups seem safe as they lounge on a Patagonian beach, slowly building their strength and preparing themselves for their departure into the sea. No large predator threatens them from the land, except for those ubiquitous hunters, human beings, and many of their beaches are, in any case, difficult to reach for they lie below steep cliffs. And on the other side of the beaches, they are protected by the sea.

But the sea is not as secure a barrier as it might seem. Every now and then, one of the tall waves thundering in from the open ocean carries within it a dark sinister presence. As the crest curls over, the cascading water reveals the huge black-and-white flank of a killer whale, travelling fast towards the shore. It surges up the beach, still supported by the water of the dwindling wave, and with one powerful beat of its great tail, it lunges into a group of unsuspecting seal pups. As they scatter, squealing with alarm, the whale seizes a struggling pup in its jaws. By now, the next wave is approaching. The whale turns broadside so that it is lifted back into the sea. As it swims away at speed, it jerks its great head upwards and sends the pup, still alive, spinning through the air. No sooner does it fall back and hit the water than the whale flicks it into the air again with a blow of its tail or seizes it with its jaws and thrashes it on the water. Why the whale should sport with the helpless pup in this way can only be guessed. Maybe such a battering loosens the hide from the little carcass. Maybe it is the jubilation of a triumphant hunter. It does not last long. Within a minute the pup is dead and swallowed.

Such swift and arbitrary death, due neither to weakness nor fault, but to sheer chance, is the fate of vast numbers of animals that every day are eaten by predators. A grasshopper, slowly chewing a leaf-blade, is suddenly struck by the clubbed end of a muscular tongue projected like a lance from the mouth of a chameleon; a field mouse in the twilight of an English wood, searching for seeds, is transfixed by the curved talons of a pouncing owl and may be dead even before its captor's beak begins to rip it apart; a lizard in the Arizona desert, stabbed by the hypodermic fangs of a rattlesnake, is paralysed as venom is injected into its veins and it can offer no resistance as the snake takes it in its jaws and swallows it head-first.

Some ants hunt in swarms many thousands strong, scouring the forest floor for centipedes and spiders, termites and scorpions, mice and lizards, indeed any small inhabitants of the rainforest floor. Ants in both Africa and South America have, independently, evolved this way of hunting. In South America, they form armies three-quarters of a million strong. You may find one temporarily encamped between the buttresses of a tree or beneath a fallen log. It has collected into a huge ball, up to a metre across. The outside surface is formed by a lacy sheet of soldiers, their legs linked together, their huge jaws agape, ready to slice into anything that might interfere with them. Within this mass, the smaller workers have in a similar fashion created chambers in which the pupae hang. And at the heart of the community sits the queen. She is two centimetres long, twice the length of any other individual in her army. Her body glistens with a special polish for she is continually groomed by her servants. Her head is large with a brain many times bigger than that a worker possesses, although she does not have much call to employ it, for her sole function is to produce eggs, and her abdomen is an immense egg factory. From it, during her lifetime, she ejects eggs in an almost constant stream. As each emerges, it is received by a worker and carried off to one of the nursery chambers, there to be carefully cherished.

Every morning, the majority of the soldiers leave the encampment and set off to hunt, forming a long brown column several centimetres wide that snakes across the forest floor. At its head, the scouts advance excitedly. Like all individuals in this army, they are blind and they search for their prey by tapping with their antennae. As they scurry forward, they rub the ground with their bodies, laying down a trail of scent that others behind will be able to follow. If the path leads over a steep bank or a log, workers will cling to one another to form a living ladder up which the rest of the column clambers. If they traverse a patch of sunshine, the soldiers link legs and form a roof over the path so that the workers, who are less well armoured, are protected from the damaging heat.

After travelling for thirty metres or so from their encampment, the soldiers at the head of the column begin to fan out and start to hunt. If they discover a grasshopper or a beetle, they swarm all over it, sinking their jaws between the joints of its external skeleton and dismembering it with surgical precision. The fragments may be eaten there and then or tucked away beside the column to be collected when the hunt is over and taken back to feed the queen and those who stayed back in camp to tend and protect her.

Day after day, the army pillages the surrounding forest. Then the eggs the queen has been laying in such abundance begin to hatch into small grubs. The pupae, too, are beginning to split and new adult workers and soldiers are emerging. Now there are many more mouths to feed. The neighbourhood has already been intensively plundered, so that night the queen stops laying and the entire community marches off, workers carrying the grubs. After it has travelled a hundred metres or so, it bivouacs. The following morning, a raiding party sets out into the new territory. Each night for the next two or three weeks, the army makes another long march through the forest until the grubs, which have been feeding voraciously, are fully grown and begin to turn into pupae. Then once again, the army makes a more permanent camp, the queen starts to produce more eggs and the month-long cycle is repeated.

Few animals can survive the sustained attack of this devastating army. Wasps have one of the most poisonous stings of all insects, but their weapons are no defence against ants. The soldiers race into a wasps' nest, driving the occupants out. Some wasps will flail about, seeking a target for their stings, but to no avail for the ants are too small. Those wasps that stay are killed. Most fly away. The nest is ripped apart and the pallid thin-skinned grubs passed back down the column for slaughter. Even large animals such as tethered dogs may be blanketed by ants and die from the accumulated shock of ten thousand bites. There is no sure defence against the ant army except to get out of its path as quickly as possible.

That, indeed, is the way in which most animals escape their predators. But there are other ways in which even the most defenceless animals can improve their chances of survival.

On the open plains of east Africa, grass-eaters – zebra, antelopes and gazelles – manage to find protection, paradoxically, among their own number. A gazelle, grazing by itself, is an easy target for a cheetah. If it is to feed at all, it has to lower its head, so losing sight of its surroundings. Those are the moments when the cheetah can creep slowly forward. As soon as the gazelle lifts its head again, the cheetah freezes. By such stealth, it can usually get within fifty metres of a solitary gazelle. If it does, it has a very good chance of catching the gazelle for it is able to reach its maximum speed within a few metres, and then it is the fastest animal on four legs. Sprint and jink though the gazelle may, there is nothing to deflect the cheetah from the chase. Relentlessly, the gazelle is overhauled. A sweep from the cheetah's foreleg trips it. A pounce, the cheetah's jaws fasten on the gazelle's throat, and it is quickly throttled to death.

But if the gazelle grazes in a herd of a hundred or so, its chances of survival are dramatically better. To begin with, it is more likely to get an early warning of the cheetah's approach, for even when its own head is down grazing, other heads are up and scanning the surrounding country ready to sound an alarm with a snort. At such a signal, the herd takes to flight. The cheetah, forced to make its run when it is still some distance from the herd, may now spend crucial moments distinguishing its chosen target from among the confusing mass of racing bodies ahead of it. And even when it succeeds, there is still a chance that after pursuing it for some way another gazelle passing in front will impede it and its first, tiring target will be able to escape. Without question, a gazelle in a herd is much safer than a gazelle by itself.

The open sea, like the open plain, offers no hiding place either and many small fish, preyed upon by shark or barracuda, dolphin or tuna, adopt the same strategy as the gazelles, seeking their safety in numbers. Herring congregate in immense shoals, a kilometre across, containing many millions of individuals. If a barracuda approaches, those on the outer margin of the shoal dart inwards, taking refuge among the silvery bodies of their companions so that the whole shoal bunches. If the barracuda charges, the herring flee away on every side, creating a clear tunnel through the shoal. If the barracuda presses its attack, then once again the great number of bodies darting in all directions make selecting a target difficult. This indeed may be one reason why animals that find protection in these great aggregations are nearly always virtually identical in appearance, irrespective of age or sex. Were a minority to have distinctive markings or shape, they would be easier to focus upon and catch. If someone throws a number of tennis balls at you, it is easier to catch a single coloured one than to catch one that is identical with all the rest.

Even animals that normally live relatively lonely lives may find it safer to congregate when faced with a particular danger. Puffins spend most of their time fishing out on the open ocean, but in spring, in order to nest and breed, they have to return to land. As many as a million arrive within a space of two or three days on the island of St Kilda in the Scottish Hebrides. With them come their prime enemies, the greater black-backed gulls. They too are going to nest and they will rely on puffin meat to feed their chicks.

The puffins nest in holes on the steep grassy cliffs. Within a burrow, they are safe from the gulls. Standing by the entrance hole they are still in little danger, for they can duck inside for refuge if necessary. Out at sea fishing, they are not easy to catch either, for they are swift in the air, moving on rapidly whirring wings. Even if a gull out-manoeuvres them, they can escape by diving below the surface of the water where the gulls cannot follow. They are, however, vulnerable when getting from one of these places to the other. Leaving their nest-holes they have some way to go before the sea can provide them with a refuge. And coming back from the sea, laden with fish for their young, they can only find safety in their nest holes. So the black-backed gulls wait for them in the air in front of the cliffs, wheeling and circling on the up-draught created as the wind, blowing in from the sea, is deflected upwards. They are quite capable of grabbing a solitary puffin in mid-air with their beaks or even beating it to the ground with blows of their wings.

The defence the puffins employ is to gather in a huge aerial wheel, a kilometre across, that circles in front of the cliffs throughout the day. Puffins leaving their homes join it immediately and travel round within it until they reach the seaward side and relative safety. Those coming in from the sea do the same thing in reverse, leaving the wheel with a sideways dive when they are within a few metres of their nest-hole. Though the gulls occasionally try to catch puffins flying within the wheel, they seldom succeed. The number and density of flying bodies make it almost impossible for them to select and catch a particular individual. Most of their victims are stragglers who, for one reason or another, fail to gain the safety of the wheel.

Not all the hunted are totally unarmed. Although few have the physical strength to repel their attackers, even a relatively feeble animal can deter aggressors if it has chemical weapons. Amphibians have moist skins and keep them so with slime produced from small glands distributed all over the body. In many species, some of these glands are modified to produce poison. The great cane toad has concentrations of them, forming wart-like lumps behind each eye. If you pick the animal up, a white milky liquid oozes out of these patches. A really big toad, if it is seriously irritated, can squirt a jet of poison and hit an attacker a metre away. If its enemy is still not daunted and picks up the toad in its teeth, the poison acts so swiftly and powerfully on the mucous membranes of the mouth that the attacker drops the toad almost immediately.

It is, of course, better for both concerned if engagements of this kind can be avoided altogether. Then the hunter does not waste its time on something it cannot eat, and the hunted saves its secretions. So many poison-producing amphibians give vivid and unmistakable warnings that they have such defences at their disposal. The fire-bellied toad normally prefers to remain hidden and its back is patterned and coloured in a way that enables it to do so by blending in with its surroundings. But if it is discovered and threatened, it twists its legs outwards and arches its back in such an extreme contortion that it suddenly and disconcertingly exposes its underside – and that is a vivid scarlet, a spectacular warning that its skin contains a burning poison.

The spiny newt of China goes through the same sort of contortions to warn off those that threaten it and adds a special deterrent all its own. If it feels compelled to release its poison, it does so instantaneously by forcing its ribs outwards with such vigour that they break its skin and rip open the poison glands.

The most lethal amphibian venom of all is secreted by tiny arrow-poison frogs that clamber about in the leaves littering the floor of the South American rainforests. They have such confidence in their defences that they make no attempt whatsoever to hide themselves. Some are vivid pink, others black and yellow, acid green or maroon with metallic blue spots. One ten thousandth of a gram of their poison is enough to kill an adult person. This very virulence, while providing excellent defence against most attackers, brings the frogs death at the hands of humans. Forest-living indigenous people catch them and roast them over a fire so that the poison drips from their skins. The people collect it in pots and smear it on the tips of their arrows and blow-pipe darts. So little is needed that one small frog two and a half centimetres long produces enough poison to arm fifty arrows.

Few mammals have chemical defences. The skunk is an exception. It has glands just beneath its tail which produce considerable quantities of a most evil-smelling liquid. You might suppose that a bad smell would not deter a really hungry hunter, but anyone who has received a full squirt from a skunk knows very well that it is almost unendurable. The stench is so powerful that you feel – and sometimes are – violently sick. If any quantity gets on your clothes, you might as well destroy them. If it gets into your eyes, you may be unable to see for several hours.

Like the frogs, the skunk does its best to prevent tiresome confrontations by displaying vivid keep-off signs. It is boldly patterned in black and white. There are several different species in the Americas, each with its own distinctive combination of stripes and blotches. All of them give you fair warning of their character by deliberately making themselves conspicuous and waving their bushy tails. The little spotted skunk puts on a particularly impressive performance. First it stamps vigorously with its front feet and erects its bristling tail. If you walk closer, it does a hand-stand, hoisting its hind legs in the air and pointing its tail over its head towards you. If that does not deter you, it drops back on all fours, turns its back on you and squirts. The jet can easily travel a couple of metres. If the wind is behind it, you can get more than enough standing twenty metres away. Nor can you dodge these attacks by approaching obliquely. The skunk has a pair of these glands and not only can it vary the strength of each jet according to the side from which you approach but it is even able to twist the nozzle of the gland so that the spray shoots out at an angle. So effective is this defence that nothing hunts skunks – and the skunks seem to know it, judging from the jaunty and self-confident way in which they bustle about their business.

To be effective, these warnings have to be understood by animals of all kinds. Amphibians need to signal to mammals, insects to birds, mammals to reptiles. The codes used have, therefore, an almost universal meaning. A bold pattern of yellow and black is one of the commonest. It is displayed not only by a species of arrow-poison frog but also by the fire salamander, moth caterpillars with stinging hairs, a little box-fish that discharges venom when attacked, a beetle that exudes a caustic liquid which blisters the skin, and bees, wasps and hornets with some of the most painful of all insect stings.

It is also the colour pattern flaunted by hover flies and clearwing moths that visit English gardens alongside bees and wasps.

But these have no stings. Their vivid colouring is a hoax. By masquerading as poisonous insects, they avoid attacks from birds that might otherwise eat them.

Their resemblance to wasps is extraordinarily close. The hover-fly only betrays its true identity, to the casual glance, by moving through the air in the jerky fashion that is characteristic of its family. The moth's physical appearance is not quite as wasp-like but it adds to its disguise by making a buzzing noise.

This tactic of drawing attention to yourself when you have no deterrent to back up your threat, may seem a recklessly hazardous one. And indeed, it is not universally successful. Some species of birds have developed the ability to distinguish between model and mimic and will feed on the imposters. Nonetheless, the strategy, on the whole, is a successful one, for such impersonations are abundant. Tiger beetles, with red and black stripes, which have formidable jaws and readily use them, are mimicked by grasshoppers; ladybirds, which have black spots on their red wing covers and poisonous blood, by cockroaches for they, as far as many birds are concerned, make good eating. A South American cricket is not only patterned like a wasp but adds a mime to give the impression that it is equally powerfully armed. It walks on only five of its legs and holds the sixth out stiffly behind it so that it appears to have a sting projecting from the tip of its abdomen. The caterpillar of a Costa Rican moth, in one of the most extraordinary of all mimicries, has a pattern at its rear end that makes it look like a tiny viper.

But most animals, seeking to avoid attack from predators, use disguise in a quite different and more cautious way. Alongside the brilliant arrow-poison frogs sit other frogs that are well-nigh invisible. Their brown coloration perfectly matches the decaying leaves around them, their blotches and lines disrupt their outlines. On beaches, vulnerable plover chicks, not yet able to fly, crouch low and motionless, the colour of their down so close to that of the pebbles around them that their major hazard comes, not from being seen and eaten, but from *not* being seen and trodden on. In the far north, where the landscape is turned white in winter by snow, mammals like rabbits and birds like ptarmigan have to change from brown to white and back again when the snows melt in spring.

Insects are the great masters of disguise. True bugs resemble thorns; butterflies, with wings closed, resemble dead leaves; moths look like patches of lichen. Such disguises, when necessary, can be enhanced by posture. Caterpillars of geometrid moths not only resemble twigs in the colour and texture of their skin, but they grasp a thin branch with their hind claspers and hold themselves up at an angle so that they look like twigs. African tree hoppers gather in groups at the top of plant stems and together look like a spike of withered flowers. One species of beetle in Brazil, when alarmed, immediately folds up its legs and flattens itself sideways, exposing its white underside and so takes on the appearance of a bird dropping. And in order that its symmetrical beetle outline should not be given away, it stretches out a white flattened front leg to one side, suggesting that the dropping in question was a rather liquid one that has splashed.

But two can play at this game. Just as the hunted may use disguises to escape from hunters, so the hunters may use them to lay an ambush.

A species of angler fish that lives in the Sargasso Sea is blotched and betassled in a way that matches the floating Sargassum weed so closely that the angler is virtually invisible to the eye of a human being, just as it is to that of a small fish, a shrimp or any other marine creature that might drift through the surface waters of that stagnant sea. Even such a perfect match would count for little if the fish had to wave its fins in order to maintain its position in the water, or if it were to move independently from the weed. That does not happen. Its front pair of fins are muscled in such a way that it can grasp the fronds of the weeds around it. When the weed sways, so does the fish.

Insects in the forests of Malaysia visiting the elegant white flowers of an orchid, may walk straight into trouble. One of the fleshy petals suddenly moves and two hooked arms shoot out from its tip. The orchid mantis's disguise is near-perfect. It has shields on its body and flanges on its legs that exactly match the hue and surface texture of the orchid petals. Neither the eye of an insect or a human is likely to detect the deception until the mantis moves. And by then, for the fly, it is too late.

Some hidden hunters go further. They bait their ambush. The death-adder in the Australian desert is so accurately matched to the colours and shapes of the gravel that it is almost impossible to detect, unless a movement draws your eye to it. And move it does. The end of its tail is thin, pink and very mobile. The snake makes it wriggle so that the apparently disembodied filament appears to be some kind of succulent worm. In Iran, the spider-tailed horned viper takes mimicry to extraordinary lengths – the end of its tail is divided and the reptile moves it about in an uncanny imitation of a large, tasty spider. A bird that thought so and decided to make a meal of it would quickly die. In South

America, a horned toad regularly attracts its prey by – almost unbelievably – waggling its toes.

Other hunters use objects as bait. A small assassin bug in Costa Rica, having ambushed a termite and sucked its body dry, holds the husk in its jaws and loiters near one of the entrances to the termite hill. When a worker, coming out of the nest, detects the corpse, its instinct is to pick it up and remove it, for that is part of its regular nest-cleaning duties. As it goes to do so, the bug stabs it with its dagger-shaped mouth-parts. Once again the assassin feeds and once again it uses the left-overs of its meal as a bait to attract another victim. A single bug may catch ten or more termites in succession in this way. In Japan, herons use bait in just the same way as human anglers. One will stalk along the margin of a lake, picking up insects dead or alive, or small pieces of bread or biscuit dropped by visitors. With a flick of its head, it throws them on to the water. There they float. Small fish soon rise to collect them and as they do, the heron neatly spears and swallows them. Some individual birds do not use edible bait, but inedible lures, such as feathers. This skill is, apparently, a recently acquired one, for initially only a small number of Japanese herons were observed doing it, but the habit is spreading. It is now being seen on ponds in the United States where fish are accustomed to being fed.

Just as camouflage is used by both sides in the contests between hunters and hunted, so is the technique of baiting. Californian skinks, handsome little lizards, have conspicuous blue tails. As one lies basking in the sun, it may be pounced upon by a hawk or even grabbed by a human being. Almost invariably, when that happens, its blue tail drops off and lies wriggling on the rock with such vigour that the aggressor's attention is diverted to it, often for sufficiently long for the skink itself to slip free and escape. It achieves this dramatic and disconcerting act of self-amputation by suddenly contracting the muscles in its tail and breaking one specially fragile vertebra in half. Having done so, it then regrows its tail, though it is not always as long as the original and internally it is quite different, for instead of bony vertebrae, it has only a tube of cartilage.

Butterflies are at risk from attack by birds. One peck on the head will kill them. But the hairstreaks have evolved a way of deflecting that fatal blow. The rear ends of their wings are elongated into long tendrils which resemble antennae and are, indeed, more conspicuous than the insect's real antennae. As they sit, they make these filaments move and flicker. Birds are certainly deceived for they regularly attack, not the butterfly's true head, but its tail. And having made the mistake a bird seldom gets a second chance, for the startled butterfly takes off, not in the direction the bird might have expected but, apparently, backwards.

So various and ingenious are the defensive techniques adopted by the hunted that, in some cases, hunters find it worth their while to work together in teams. Cormorants are usually independent fishermen. They dive beneath the surface and pursue fish with their eyes open, their wings clasped close to their bodies, beating their feet powerfully. But on the Amazon in Brazil, the birds collaborate. They gather together in a dense fleet a thousand or more strong, and paddle steadily towards an inlet or a small bay, splashing vigorously with their wings and feet. The panic-stricken fish bolt ahead of them until a whole shoal has been herded together and trapped between the birds and the shore. When the water is sufficiently shallow, the cormorants break ranks and start to dive, gathering great quantities of the fish with huge commotion.

Pelicans have developed an even more accurately coordinated technique. Several dozen of them will gather together in a squadron. As they paddle sedately forward across the water, they suddenly form a ring. Then with the precision of a corps de ballet, they simultaneously dip their heads into the water. If a fish, swimming within their circle, manages to escape from one baggy bill, it is likely to swim straight into another.

Even fish have discovered how to work in teams. Striped marlin, one of the most ferocious and swiftest of hunters, often operate in groups of three or four. When they discover a shoal of smaller fish, they will harry them on all sides, driving them into such a dense concentration that the technique has been called 'meat-balling'.

Among mammals, lions habitually hunt in teams. Individuals chase and pounce on their prey if a suitable opportunity occurs, but they have substantially greater chances of success if they work together. This is usually done by the lionesses. A group of them, maybe as many as half a dozen, will slowly get to their feet from where they have been lying with the rest of the pride and, leaving the cubs and the males behind, walk off in a fashion which, although leisurely, has a grimly purposeful air. They know the country around their home well and, doubtless, the habits of the game animals on which they prey.

As they approach a grazing herd of, say, wildebeest, they spread out in line abreast and begin the stalk. Like the cheetah and other hunting cats, they crouch and freeze motionless if their quarry looks in their direction, moving only when they are unobserved. Those at either end of the line usually advance rather faster than those in the centre so that a pincer movement develops. They certainly pay close attention to one another's progress, frequently glancing from side to side to check on each other's position.

They are not as swift as cheetah and have to get considerably closer, usually to within twenty metres of their target, if they are to have a reasonable chance of overtaking it when the race begins. As they approach this critical distance, either they are detected or one of them charges. In either case, the herd bolts and the lions sprint towards them. The lionesses on the flanks may well drive some of the wildebeest towards their companions in the centre of the line. Whether they do so deliberately is a question that is still being debated. Most lion-watchers have come to the conclusion that the effect, when it happens, is a fortuitous one and that there is no predetermined strategy by which one or more lionesses take the job of driving while others deliberately wait in ambush. Each animal reacts individually to the movements of the prey and each benefits from the fact that her companions are doing the same thing.

Hunting dogs in Africa and wolves in North America also hunt in teams, snapping at the heels of an antelope or a moose, one taking over from the other until their victim is so exhausted that they can get a grip on it with their teeth and pull it down. But as in a team of lionesses, it seems that there is no specialisation of role. One particular wolf or hunting dog is not always the individual that takes the first grip on its victim. In only one non-human species has such specialisation of role within a hunting team been demonstrated, and that is in our closest relative, the chimpanzee.

In the Kenyan savannahs where chimpanzees were first studied intensively in the wild, the animals live primarily on a vegetarian diet of fruit and leaves, together with insects such as termites. On rare occasions, these chimpanzees have been seen to kill and eat other animals such as young baboons and bushbuck. Sometimes the capture is made by an individual without help from others. Sometimes several chimpanzees may take part in the hunt, but they do not seem to collaborate to any greater extent than lions or wolves. But a ten-year study in the West African forest of the Ivory Coast revealed that chimpanzees there hunt regularly and do so in teams within which there are specialised roles, habitually taken by particular individuals. Since it seems that the rainforest rather than the savannah was the chimpanzee's original home, hunting should therefore be regarded as being typical of the animal.

The forest of the Ivory Coast is thick. The chimpanzees live in groups of sixty or so, though they move in much smaller parties. Individuals may leave one party and travel with another. Sometimes the parties will join. At other times, the animals will disperse widely. The parties communicate regularly with one another by hooting calls and, most dramatically, drumming. This is a complicated gymnastic performance. A male, after some preliminary hoots, suddenly takes a flying leap at one of the huge plank-like buttresses that radiate like fins around the base of some tall forest trees. He grabs the top edge with his hands, thumps the face of the buttress several times with his feet and then does the same thing on the other side of the buttress with his hands before springing down to the ground. The whole performance is enacted in one continuous movement and is followed by a series of screams. At a distance, the leafy forest drowns the higher-pitched calls and all you hear are the rapid drum-beats. It is easy to think that they are made by men.

Like the savannah chimpanzees, these forest-living animals eat fruit, leaves and nuts. But at least once a week, they hunt for meat. During the two months of the rainy season, they may do so every day. Their prey are monkeys, principally two species of colobus, the red and the black-and-white, both of which are abundant in the forest. A colobus weighs less than half a chimpanzee, so they can venture out on to branches that would break under a chimpanzee's weight. They are also excellent leapers, capable of jumping from one tree to another. Chimpanzees, on the other hand, usually only cross if the branches are so close that they can swing over. Even if they are forced to jump, they cannot cover anything like the distances achieved by the colobus. So in theory, a colobus should find it easy to escape from a chimpanzee. The chimpanzees can only catch them regularly by working in teams.

The hunters are the half dozen or so experienced adult males in the group. Between them, they have to take four quite different roles. The driver is responsible for getting the troop of colobus moving through the canopy. He may be the youngest of the team, sometimes even a young adolescent. He does not chase the monkeys, but only keeps them from settling. The blockers – and there may be several of them – must take up conspicuous positions in the branches on either side of the drive, so preventing the monkeys from breaking out. The chasers join in the hunt once the monkeys are on the move. They must spring up the trees as the monkeys are driven into them and they are the ones that usually make the kill. And finally there is the most skilled job of all that requires the most experience and judgement, the ambusher. He is usually an old male who can anticipate which way the colobus will go and climbs a tree well ahead of them, so completing their encirclement; in any one team, he is usually the same individual.

Before a hunt, the team assembles gradually. It may be that the drummings by the males have served to communicate not only where each one of them is, but what mood they are in. At any rate, the males leave their parties and come together in a posse. The change in their behaviour is dramatic. There is no more calling and hooting, no picking up of fruit or plucking of leaves. They pace together through the forest in silence, scanning the canopy intently, sometimes stopping and listening for the calls of colobus. It may take only twenty minutes or as long as two hours before they find the monkeys and are sufficiently close to them to launch an attack. Suddenly, the driver runs up a tree, climbing swiftly, hand over hand. He will, if he can, isolate one or two monkeys from the main troop. Most of the chimpanzees stay on the ground as spectators. The adult females bob and dance with excitement, standing upright, craning their heads back and forth to see just what is going on. If one monkey is separated, the blockers dash up into the trees ahead to take up their positions, crashing through the branches in a way that is quite unlike their normal movements.

Now all is action. The ambusher sprints ahead to find the place where he will hide in the leaves, while the chasers move in front of the driver and run along the branches trying to grab the monkey and chasing it towards the place where the ambusher sits hidden. The colobus, driven forward between the blockers, is deceived into thinking that an avenue of escape lies ahead until suddenly the ambusher reveals himself. The monkey hesitates, turns back and is grabbed by the catchers. As they do so, they scream with excitement. Their calls are immediately taken up by the whole team and the spectators on the ground so that the forest rings with wild and terrifying shrieks.

More than half these hunts are successful. Some only last a few minutes. If a particular monkey is chased and harried for as long as ten minutes, it may become so stressed that eventually it gives up trying to escape and sits to face its death without screaming or even resisting when the hunters finally seize it. Then it is torn limb from limb up in the tree. Sometimes, it is taken to the ground. There a scrum of excited adults, both male and female, surround it. Two of the senior males of the group, whether or not they have taken part in the hunt, tear the body in two. Each is then surrounded by other adults from the group. In order of seniority, they are handed fragments or allowed to tear off pieces. If the body is a small one, the younger hunters may not be given a share. Adolescents and babes never get anything.

In the distance, the bereaved colobus still sound their alarm calls. The chimpanzees, gnawing on the joints, ripping muscles from bone, occasionally squeal in irritation as they squabble, but for the most part, after the excited rushes and the yells of triumph, there is contentment. A human observer may find the scene horrifying. The limp body of the monkey is only too human in its proportions, the cries of triumph only too reminiscent of the yells of a human mob bent on violence. Some of us may see in these bloodied simian faces the image of our own hunting ancestors. But if we do, we should also discern in them the origins of the teamwork and collaboration that we have brought to an unparalleled peak of complexity and that has brought us some of our greatest achievements.

FIVE

Finding the Way

The end of the day on the plains of East Africa is, by popular acclaim, a time of great splendour and beauty. The sun, a huge scarlet disc, drops with dramatic speed behind rafts of clouds, staining them first gold and then red. A hush falls over the landscape as the animals prepare for the coming of the night. The herds of antelope cluster together a little more tightly; vultures and storks alight in the tops of trees to roost; baboons clamber up into the branches where they will be safe from prowling leopards. But if you are travelling on foot, and particularly if you are without a light, this can be a time of real alarm. As darkness spreads swiftly across the land – far more quickly than those of us who live in high latitudes are used to, for there is virtually no twilight – the track you have been following disappears into the gloom. Soon the only things to guide you are the silhouettes of trees or rocks, black against an only slightly less black sky. Unseen thorns begin to snatch at your clothes and rip your skin. You can see so little as you blunder on that you are an easy target for any animal seeking fresh meat. If you keep walking, and if the moon has not yet risen, you are likely very soon to be lost.

But not all mammals are as dependent upon their eyes as you. In the thorn trees, bush babies, small furry primates, are emerging from their holes in the tree trunks. Their large eyes are much more sensitive than yours, but they do not rely on them entirely. They have another sense to help them find their way through the blackness.

As a male sets out along a branch, he leans over sideways, lifts one of his back legs and carefully expels a few drops of urine on to the sole of his foot. Then he reaches back with his hand and rubs his foot with it. That done, he leans over to the other side and does the same thing with his other hand and foot so that both his palms and his soles are anointed with his pungent urine. From now on, as he goes about his nocturnal perambulations, he leaves a smelly trail behind him. This warns any wandering bush baby from another group that this territory in the tree already has owners who will defend it. It also proclaims his sex and particular identity. Similar marks left by females leave similar messages including information about their sexual receptivity. But the long smelly trails snaking along the branches and renewed repeatedly every night, also mark out the main highways through the tree so that, if necessary, the bush babies can scamper along them at considerable speed in the pitch blackness.

On the plains, small rodents that have spent the daylight hours in the safety of their burrows are venturing out under the cover of night to look for food. They too have a network of trails winding through the sparse grass. Sometimes these are long tunnels running beneath withered drooping stems, sometimes no more than faint dust-free tracks crossing patches of bare earth.

They too are given a special smelly identity, for their owners have scent glands in the soles of their feet.

Hyaenas, having spent the day skulking aimlessly in the thickets or strolling in a leisurely manner across open ground, now become transformed into determined hunters. As they set off across their territory in the darkness to look for prey, they leave pungent signposts. Not only do they defecate communally in special latrine areas, but they 'paste'. Special glands beneath the tail produce a powerfully smelling white substance. When they reach the borders of their territory, they take a few steps with lowered hindquarters so that long grass passing between their back legs rubs against this gland and acquires a smell that even a human nose can easily detect.

A hyaena's nose, however, is vastly more sensitive than ours. We have smell-detecting membranes in our noses that are about the size of a large postage stamp. A hyaena's nasal membranes have a surface area seventy times bigger and the richness of the information they can gather is so great and varied that it is difficult for us to appreciate it. With a sniff a hyaena can perceive not only the here and now but, simultaneously, a whole series of events stretching back into the past. It can identify the brief passage of an animal that might have run across the ground in front of it several hours previously. It can recognise the particular smell signature of each member of its group. The pasted patches of grass must shine in the distance like lighthouses and the pack's trails, perfumed by their paws, must stretch ahead like lines of reflector studs down the middle of a motorway. For mammals with such receptive and informative noses as hyaenas, bush babies and mice, darkness is little impediment in finding their way around.

Birds were long thought to have a comparatively feeble sense of smell, but this turns out not to be true – they have as many

genes coding for olfactory receptors as we do. Turkey vultures detect the smell of dead animals from high in the sky. One group in particular uses olfaction in deep-sea navigation. Shearwaters and petrels belong to a group called the tube-noses, for their nostrils are protected by a pair of tubes which extend some way down the beak from its base and these give them a much better perception of smell than other birds have. In the open ocean they use this to detect a chemical released by phytoplankton that are being grazed upon by krill which in turn are eaten by the birds' prey – they can use this signal to locate their food. In a terrible twist, this same compound is also released by plastic that litters the ocean, leading some species of seabirds to swallow this indigestible material and even to feed it to their young, with tragic consequences.

Each year, such ocean-going birds have to come ashore to breed. They assemble in colonies many thousand strong and nest in long holes that they have either taken over from rabbits or dug for themselves.

By day, a colony of petrels or shearwaters is a quiet, apparently deserted place. Some of the adults are at sea feeding. The rest are out of sight within the holes, incubating their eggs or brooding their young. There they are safe from the gulls or skuas that harry other nesting sea-birds. The adults extract oil from the sea creatures they eat and then regurgitate it for their chicks. That in itself may be a messy and therefore a smelly business. Most species also squirt the oil at intruders as a form of defence so that the ground around the nest becomes impregnated with it. This combined with the smell of their droppings and the musky odour of the birds themselves, makes such colonies very smelly places and enables the birds to use the smell to guide themselves back through the darkness of night.

A Manx shearwater colony has a particularly powerful stench. The birds, when they come shooting in through the darkness, seldom land exactly beside their nest hole. More usually, they come down with a thump a metre or so away. They are then faced with dozens of holes closely packed together. Identifying theirs from those that surround it would be a baffling task in the darkness without the ability of the bird to recognise the particular smell of its own burrow.

One or two other species of bird have developed a much more accurate technique of finding their way in the dark. Swiftlets, tiny birds that live in South-east Asia and Australia, nest in caves. Nocturnal birds, such as owls which depend upon their hyper-sensitive eyes to fly in the dimmest of lights, might manage to fly in parts of the cave near the entrance, but some species of swiftlet nest in chambers so deep that no light whatever reaches them. There owls would be completely helpless and grounded. But the swiftlets fly fearlessly and unerringly through the blackness for they have another sense by which to steer.

As they enter their caves, they start producing a series of high-pitched clicks. The frequency with which they make them varies. In the spaciousness of a large chamber they may make only four or five a second, but as they approach the rock walls and need to know exactly where they are in order not to crash into them, they increase the speed of the clicks until they are emitting as many as twenty a second and the sound becomes, to our ears, an almost continuous rattle. The time taken by the sound of each click to bounce back from the rock to the bird enables the swiftlet to judge just how far away it is from the rock wall ahead.

For this system to work, the click must be so short that its echo is heard before the next click is made and drowns it. The bird must also know in which direction the echo-producing obstacle lies. To do that, it assesses the difference in intensity of the sound in each ear and the infinitesimal difference in time that it takes to reach one ear before the other.

Clearly all this demands apparatus of great sensitivity. But the swiftlet's technique is crude compared to that developed by bats. They have refined every aspect of the system. First, the pitch of the sound. The higher it is, the smaller the surface its echo can reveal. We can hear some bat sounds, particularly when we are young and our ears are sharp, but these sounds are the bats' social utterances. Those they use for navigation are so high-pitched that they are ultrasonic, far beyond the range of any human ear. Some are so high that they enable their makers to detect the presence of a wire no thicker than a human hair stretched across their flyway.

Second, the intensity of the sound. The louder it is, the more distant the object it can detect and the bats produce clicks which, if translated into frequencies that we can hear, would sound as loud as a pneumatic drill. This, however, causes a major complication. It is so loud that were the bats to hear it, their hyper-sensitive ears, tuned to detect the faintest of echoes, would be seriously overloaded. This problem is dealt with by a muscle in the middle ear attached to one of the trio of tiny bones that transmits the vibrations of the ear-drum to the tubular organ in the skull that converts them into nervous stimuli. As each click is made, this muscle pulls aside the bone so that the ear-drum is momentarily disconnected. It is then replaced in time to receive the echo. And this it may do more than a hundred times a second in perfect synchrony with the clicks of the calls.

Third, the faster the clicks are emitted, the more up-to-date information the bat will receive as it negotiates the obstacles in the cave and dodges through the branches and creepers of the night forest. Some bats can send out a stream of two hundred clicks in a single second, each lasting only a thousandth of a second and spaced sufficiently from the others to allow each echo to be heard.

Echo-location, or sonar as it is called, is only used by small insect-eating bats. Nearly all the much bigger fruit bats rely on their large eyes to find their way and in consequence, like owls, they cannot fly in caves where there is no light whatsoever. The insect-eaters, however, get such an accurate picture of their surroundings from their sonar that they hardly use their eyes at all. Indeed, these have become so tiny that they are of little practical use when flying.

There is another dark world where animals use sonar to find their way around. Many great rivers – the Ganges and the Indus, the Amazon and the Yangtze – are so muddy that the animals swimming in them cannot see more than a few centimetres ahead. A dolphin is able to produce sonar clicks by forcing air through special passages and sinuses in its head. These are focused into a beam by an oval, fat-packed organ, the melon, which forms a bulge on the dolphin's forehead. This is, in effect, a sound lens and it produces a sonic-searchlight with which the animal scans the water ahead of it. A dolphin can emit up to 700 clicks a second with this apparatus and from them is able to detect not only the presence of a solid object in the water but to deduce what sort of object it is. It can distinguish between a tin full of water and one that is empty, between rock and flesh. In the open sea, dolphins use this sense to find and catch fish, but in the turbid waters of rivers, freshwater dolphins use their sonar for navigation. Because

it is so difficult to see in this dark environment, their eyes, like those of insectivorous bats, have become very small. Those of the Indus and Ganges dolphins even lack lenses, so that their eyes can tell them little more than the difference between light and darkness, between day and night. In an amazing example of convergent evolution, it turns out that the same genes have been used repeatedly in the separate evolution of echo-location in different mammal lineages, such as bats and dolphins.

Fish in these muddy rivers also have problems in finding their way. All fish can detect the presence of objects near them even if they cannot see them. A long fluid-filled tube runs beneath the skin along the middle of each side of their body. This is connected to the surrounding water by pores. As the fish moves, so solid objects in the water nearby create slightly greater pressure in the water and this the fish is able to sense by means of its lateral-line system. But many fish in these particularly turbid waters have additional navigational devices. Some have fleshy whiskers half as long as their bodies which they project forward and wave about, like a blind person with a stick. These whiskers are sensitive not only to touch but to taste, so the fish can tell what on the river bed in front of its mouth is worth eating and what is not. Catfish, more than any fish, grow such whiskers in a most spectacular way, some species having as many as six sprouting from their upper and lower jaws.

A few fish, however, in both the Amazon and the Congo, have developed a sensing system that is used by no other group of animals. They use electricity. All living organisms produce electrical impulses on an infinitesimal scale. They are the

medium by which messages are sent along the nerves; they are discharged whenever muscles contract. But these muddy-water fish have developed banks of modified muscle tissue that generate electrical charges on a much greater scale.

The Amazon knife-fish is about twenty centimetres long and very odd-looking. It has no tail fin like those of other fish, merely a fleshy stump. Nor does it have a dorsal fin on its back. Instead it has one long ribbon-like fin that runs along its underside from close to its stumpy tail to its head. By undulating this, it drives itself forwards – or, with equal ease, backwards. Above this fin, buried beneath the skin, is a line of organs which emit a stream of electrical pulses. Their voltage is very low – a mere three to ten volts of direct current – but the frequency of the pulses is very high, around three hundred a second. In uncluttered water, these discharges create a symmetrical electric field around the fish which it can sense with a series of receptors in its skin. A solid object, whether it is a rock, a fish or even a plant stem distorts this field and the fish can immediately sense the change. It is as aware of objects behind it as in front of it, and when alarmed it can reverse backwards into its hole with a speed and accuracy that any motorist would envy.

There is one limitation. If the fish were to bend its body as many fish do when they swim, the pattern of the electrical field would distort. So all the fish that use this system, whether in the rivers of West Africa or in South America, keep their bodies as stiff as ramrods and have to propel themselves by undulation of their fins. The electrical field also becomes confused if two fish using signals on the same frequency happen to meet, for one will interfere with the other. When that happens, both fish immediately stop transmitting and then start up again with slightly different pulse rates.

The capacity to generate electricity has been taken even farther by one fish in the Amazon, the electric eel. Like the knife-fish, it navigates by producing steady low voltage emissions, but it can also, with a different set of generators, produce massive shocks. These it uses to stun its prey and the discharge is so powerful that it is quite sufficient to knock over a horse standing in the shallows.

So by exploiting touch, taste and smell, by developing special techniques of echo-location and electrical sensitivity, many animals find their way around in the darkness with great precision. But for many others, including ourselves, the dawn and the return of the sun is a welcome relief from a time of enforced inactivity. At last, we and they can use our eyes again and see where we are going. Birds leave their roosts and take to the skies; monkeys jump away through the branches to find breakfast, and antelope disperse once more over the open plains to graze, knowing that if danger comes they stand a good chance of seeing it before it gets lethally close. Everyone sets off once more on their regular journeys around their familiar home ranges.

Few wander at random. Nearly all have favourite places where they regularly sleep or drink or hunt and most move around along particular trails. The African elephant shrew, a highly-strung insect-eating mammal the size of a mouse with a nose elongated into a mobile trunk, depends for its safety on knowing its trails better than any hunter that might chase it. It must be able to run full-tilt down any of its tracks, anticipating every hazard on the surface that might trip it up and leaning into familiar bends like an experienced racing driver going round a well-practised circuit. So first thing every morning, the shrew trots round the course, clearing away any twigs or leaves that might have fallen across it with side-swipes of its delicate fore-legs. Its knowledge of its home range, however, is not restricted to these tracks, even though it seldom leaves them. In an emergency, it will take a short cut and plunge across a patch of land that it never normally enters in order to reach the safely of a bolt-hole. This must mean that it is able to visualise the relationship in space of all these paths to one another. It must have, in its memory, a map.

Such a mental picture, the more remarkable the more one considers it, is probably within the minds of most animals. Gobies, small fish that can often be found in tidal rock pools, prove in a particularly convincing way that they have such a thing. One kind, Bathygobius, has the habit of leaping from one pool to another as the tide retreats. Although much of the area around a pool at this time may be exposed rock, the gobies never land on it. They know just where other pools lie and are able to judge their leaps with such accuracy that they always drop into one. Presumably they acquire this knowledge when the tide is high and they can swim from one basin to another, and they are able to translate that information into a mental picture of their entire territory.

But not all animals, even on their home ground, can rely on such a well-learned map. In the Sahara there are great areas of sand where there are no permanent landmarks from which to construct one. The sand is so hot and dry that in many places no bush or clump of grass can grow to provide a signpost from which to take bearings. Tracks get covered by wind-blown sand in minutes. Scent trails are baked dry and odourless by the fierce sun. This unpromising, bewildering country is the home of tiny ants, Cataglyphis. They live in nests below ground where they are safe from sand lizards and insectivorous birds and that is where they stay during the mornings. But as mid-day approaches, it becomes so hot that the lizards and birds retreat into the shade wherever they can find it. Now, for an hour or so, Cataglyphis can forage in safety. Hundreds of them – all female, as in all worker ants – suddenly erupt from a tiny hole in the sand and start sprinting across the dune looking for the bodies of insects that may have collapsed from heat stress. Each one searches along a zig-zag course. Every few seconds, she stops, executes a pirou-ette with her head lifted and then dashes off in another direction. Eventually, with luck, she finds and picks up a tiny insect corpse. Now she must get back to her nest as quickly as possible before she too collapses in the heat.

She does not retrace the zig-zags of her outward journey. Instead she runs in a straight line directly and accurately back to her nest-hole which may be as much as a hundred and thirty metres away. She has measured and remembered the distance she ran on each stage of its outward journey by measuring the number of strides her tiny legs have taken. Every time she lifted her head and pirouetted, she registered the new direction she was taking in relation to the sun. All this information, compiled during a journey that may have lasted as much as a quarter of an hour, enabled her to deduce the exact course she had to take in order to arrive back at her nest-hole. Incredible though this may seem, it has been proven experimentally that this is the method she uses. Individual ants have been repeatedly followed with a trolley carrying a mirror that displaces the image of the sun as seen by the ant. Individual ants, misled in this way, fail to reach their holes but go to a point in the desert that is displaced by just the amount that the sun's image was shifted by the mirror.

A worker bee also uses the sun in a similar way. Having found a group of flowers bearing nectar, she is able, taking her cue from the sun, to fly straight back along a reverse bearing to her hive. What is more, she is able to tell others what direction they must fly in order to get food themselves. She performs a special dance, walking in a circle which she then bisects while vigorously waggling her abdomen. If the dance takes place on a horizontal plane, then this waggled line will point directly at the food source. That would be remarkable enough. But the dances are normally performed on a comb in the hive where they hang vertically. Here the bees use a convention they all accept, just as we all accept that our maps are orientated with their tops to the north. In the bees' case, the vertical is understood to point towards the sun and the angle the waggle-line makes with the vertical represents the bearing, to the right or the left, along which the food source lies. Furthermore, the intensity of the waggle indicates how far along that line the food will be found.

The disadvantage of the sun as a signpost is, obviously, that it moves. Most of the journeys made by honey bees or Cataglyphis are so brief that its movement is not of great significance. When necessary, however, bees are able to compensate for it. Having found a food-source in the evening, they will fly straight back to it the following morning, still guiding themselves by the sun even though it is now in the east and not the west.

The bee dance has turned out to be more complex than was originally thought. At different latitudes the sun rises to different heights in the sky; bees from different locations therefore use a slightly different version of the dance – an accent, if you will – altering its angle to respond to the changes in the sun's position. And by making tiny robot bees, researchers have shown that the production of sound during the dance is essential. They have also shown that while dancing the bees release chemical signals that attract other bees to the dance floor, encouraging them to pay attention to the signal from their nest-mate.

Despite the success of the bees' sun-based waggle dance, the earth's magnetic field is a much more reliable guide. It is never shrouded by clouds, it does not disappear at night nor does it move. It is this unvarying ubiquitous signal that we ourselves use, of course, when we take our bearings with a compass. Birds use it too.

The homing pigeon is extremely skilled at doing so. The wild form of this bird, the rock dove, is not a great traveller and spends its life within quite a small home range. But human beings discovered a long time ago that if the birds were taken away from their homes, they had an uncanny ability to return even across many kilometres of territory that was totally unknown to them. The Roman emperor Nero, two thousand years ago, used pigeons to send back the results of the Imperial Games to his friends and relations.

Modern breeds of these homing pigeons return so reliably and are so tame and amenable that they make excellent subjects with which to investigate the homing ability. Careful experiments have made it clear that they take note of the geographical features below them and of the smell of the area around their home and when leaving their home loft they circle above it as if refreshing their memory with one last look before departing. To test whether recognising the land beneath is essential to them in finding the way, opaque contact lenses were fitted to the eyes of some, which prevented them from seeing more than a few metres ahead. The birds still found their way home. There is good evidence that they take note of the position of the sun, but birds released on an overcast winter's day when the sun was invisible still returned. But if, on days when they cannot see the sun, small magnets strong enough to drown the earth's comparatively feeble magnetic signals are attached to their heads, they get lost. The earth's magnetism, it seems, must on occasion be their guide. But how do they perceive it? That is still not fully understood, but tiny particles of magnetic material have been discovered both in their skulls and their neck muscles. Maybe it is these that enable them to feel within their bodies the contours of the earth's magnetic forces.

Many animals need to make long journeys over unfamiliar country during the normal course of their lives. Some fish migrate every year over great distances. The Atlantic salmon spawns in the rivers of Europe. The rate at which the hatchlings grow varies considerably. In the cold rivers of Scandinavia, where food is in short supply, it may take them as much as six or seven years to reach a length of ten centimetres. In southern England, they may do so within a single year. But when they have reached this size, they start to travel downstream. The journey is a slow one, for the tiny fish do little more than allow themselves to be carried by the river current. To begin with, they travel only at night and often they may not go much more than a couple of kilometres. Eventually, after several weeks they reach the sea and there they begin swimming in a more determined way, seeking food. After several years, when they have grown to full size, they start to swim back to the rivers in order to spawn.

Their outward journey was comparatively easy. The return is full of obstacles. They have to swim against the flow of the river. They may have to leap up waterfalls. Yet, with very few exceptions, they succeed in getting back to exactly the same stretch of river where they hatched. To achieve this, they use an extraordinarily refined sense of smell. The nostrils of a fish are not involved in any way with breathing but are U-shaped tubes that contain receptor cells capable of sensing the smell of water. Each river has its own unique mixture of dissolved minerals, decayed vegetation and the taint of its particular community of animal inhabitants. The salmon can recognise this cocktail first in a generalised way in the brackish water of an estuary and then with increasing precision as they follow it into smaller and smaller tributaries until at last they reach the shallows where it exactly matches the prescription demanded by their nostrils. Only then do they settle down to spawn.

Spiny lobsters spawn on the coral reefs off the Florida coast and around the Bahamas. But when the first storm of autumn stirs the water, they leave their holes in the reef and assemble in large groups. Then they form up into single file with as many as fifty in a column, each animal touching the rear end of the one ahead with its stick-like antennae, and set off briskly across the sandy sea floor, heading for deeper water. Down there they will escape the buffetings of the storms to come and there too, where temperatures are much lower, their bodily process will slow down and so use less energy at a time in the year when there is little for these animals to eat. Their navigational system may be simply an urge to move continuously into water that is minimally cooler, and they may also be able to orientate themselves from the direction of the wave surge and the pattern of ripples on the water-surface. Travelling in queues reduces the drag of the water on any one individual, except for the leader. It also gives them protection as they venture across the open plains of sand, where there are no hiding places. If they are attacked by one of their main enemies, trigger fish, the column breaks and they form circles, antennae outwards, like a group of soldiers faced with a surrounding enemy.

A similar urge to move to lower temperatures is probably the main cue for the bogong moth of Australia when it sets off on its migration. The caterpillars in the spring feed on the grassy pastures of southern Queensland and New South Wales. As the year warms into summer, they pupate and turn into small greyish-black moths. Instead of enduring the summer's baking heat, they set off on a long journey up into the Australian Alps. Every year they take exactly the same route as previous generations, for all are influenced by exactly the same topography. Doggedly, they fly higher and higher up the flanks of the mountains towards piles of immense granite boulders that lie close to the summits. And there they disappear into crevices to roost and rest.

There are not many suitable sites for this. The first-comers naturally claim the best positions, deep in the cool darkness. They pack tightly together until they cover the stone as closely as tiles on a roof. Before long, the only vacancies are near the entrance. Those moths that settle there will probably only stay for a day and then continue higher still. Some may not find suitable lodging until they get around 450 metres up in the mountain. There they spend the summer in a state of suspended animation, sustained by the fat reserves that they accumulated when they were caterpillars.

Annual changes in the weather are also the stimulus for millions of birds to migrate. Each autumn, almost half the species that breed in northern Europe start to move south. The rich harvest of insects and frogs, fruit and small rodents which supported them and their nestlings, is coming to an end. Temperatures are falling.

Some of the migrants may only go as far as southern Europe. Others will continue across the Mediterranean, over the Sahara and on into southern Africa. In the Americas, birds make similar journeys in order to escape the privations of winter. Tiny hummingbirds fly down from New England into Louisiana and then take off on a non-stop sea-crossing of eight hundred kilometres across the Bay of Mexico to the Yucatan Peninsula and the warm jungles of Central and South America. The longest journey of all is made by the Arctic tern. It may breed well north of the Arctic Circle. When the chicks are reared, some go down the west coast of the Americas all the way to Patagonia. Others fly over western Europe, the west coast of Africa to the Cape of Good Hope. Then many continue across the Antarctic Ocean towards the South Pole where they find continuous daylight while their northern breeding grounds are covered by continuous night. The journey, by either route, is at least twenty thousand kilometres and the birds fly non-stop, sustaining themselves by diving into the sea for fish as they go.

To make such journeys, birds use almost every sense we know of and probably some we have not yet identified. Some birds certainly use their eyes to follow the geographical features they see below them and have a mental map by which they steer. Many follow obvious features such as coastlines, ranges of mountains or deep valleys. In this way, many birds ensure that they cross the Mediterranean at its narrowest point, the Strait of Gibraltar, or circumvent the sea altogether by travelling east into Asia across the Bosphorus and then down along the eastern shore of the sea by way of Israel.

Swans flying down from Siberia to Britain travel in family parties so the young cygnets compile such maps as they follow their parents. They also learn the position of the wetlands that provide them with crucial staging posts where they can feed and rest before starting on the next part of their journey. This parental guidance, however, cannot always be provided. Consider the young cuckoo. It has been abandoned by its parents before it even hatched. But it too manages to find its way down to southern Africa. It must have inherited its mental map.

Some migrating birds navigate by the sun, just as bees and the Cataglyphis ant do. This has been demonstrated by experiments with captive starlings, using the technique of displacing the sun's image with mirrors. Many small birds travel at night when they are safe from attacks by hawks. They cannot therefore use the sun. Instead they guide themselves by the stars, as has also been verified by experiment. This accounts for the fact that on cloudy nights, when there are no stars to be seen, they tend to wander off course and may even get completely lost. Other species that fly through both day and night must use both methods. And without doubt, many birds as well as pigeons, are able to guide themselves by the earth's magnetism.

But how did such migrants ever learn that there was better weather at the other end of the globe or that hundreds of kilometres north of the African savannahs, for a few months, abundant food was to be had. The answer seems to lie in the past. At the end of the last Ice Age, some eleven thousand years ago, glaciers stretched across the middle of Europe and African birds had little difficulty in visiting their southern edges where, during the summer, there were rich flushes of insects and other foods but only a small resident population cropping them. As the Ice Age passed and the earth warmed, the glaciers, year by year, retreated. But still the habit of flying north for the summer persisted among the birds and it has remained to this day, even though the journey is no longer a few kilometres, but more than a thousand.

Migratory habits can alter with changes in climate and have important evolutionary consequences, as can be seen in the example of the blackcap, a small bird that nests in central Europe, sometimes visits the UK in the summer and has traditionally overwintered in Spain. Around thirty years ago, birdwatchers noticed small winter populations of the birds in the United Kingdom, that had remained through the newly-clement UK winters instead of migrating south. This has led to the development of two separate populations of blackcap, one of which overwinters in Spain, the other in the UK. Intriguingly, there are now strong signs of differences in the morphology and genes between these two populations, which arrive back at their breeding grounds at different times and as a result tend to mate with their own group, rather than the other population. In the long run, this could lead to the development of two separate species.

Perhaps the most mysterious and complex of all navigational feats begins in the Sargasso Sea, that near-stagnant patch of warm water in the western Atlantic between Bermuda and the West Indies. There, at depths between 400 and 750 metres and in temperatures of around 20°C, the critically endangered European eels lay their eggs. These develop into tiny animals so unlike their parents that the connection between adults and offspring was not recognised until the last century.

The young, called elvers, are transparent, shaped like the long slender leaves of a willow and without fins except for an undulating fringe around their margin.

These strange creatures are carried eastwards at a depth of about 200 metres by the great ocean-wide current known as the Gulf Stream. It is usually said that they are quite passive at this stage. Certainly the Gulf Stream is quite strong enough to carry them across the Atlantic to the shores of Europe. It will transport an inanimate object like a log from one side of the ocean to the other in about ten months. The odd thing is that the larval eels take around a year and a half to make the passage.

When they reach the edge of the continental shelf, in places several hundred kilometres from the European coast, they begin to change. Their leaf-shaped body narrows. They shrink a little in length and grow pectoral fins. Soon they resemble a small adult eel, except that they are still transparent. In this form, they advance on the coast of Europe. They do not feed on the way. Some swim into the Baltic, others pass through the Strait of Gibraltar into the Mediterranean and may even travel as far as the Black Sea. And great numbers move into the estuaries of Britain and western Europe.

They now have an urge to seek fresh water and start swimming up the rivers. Many males do not go far and remain in the lower reaches. The females, however, continue upstream a considerable distance. Some may even reach the higher valleys of the Alps. They use the river banks as a guide and nearly always keep within a metre of them, so avoiding the main force of the river's current. They circumvent waterfalls by wriggling through the sodden vegetation on the banks. When they enter lakes, their sensitivity to the slightest movement in the water enables them to swim straight through into the feeder rivers.

After a few months in fresh water, they begin to eat again and grow. Their bodies become pigmented and opaque, yellow on their back and sides. For the next few years they stay in these fresh waters, but even at this stage in their lives, their wanderings do not cease. As winter comes and the mountain streams get cold, the yellow eels descend to lower, warmer stretches of the rivers. When spring returns, they ascend again. Their powers of navigation remain extraordinary. Yellow eels have been caught in a Scandinavian estuary, tagged and released in another over a hundred and fifty kilometres away. They reappeared in their original river within weeks. Others have been caught and left on the ground several hundred metres from a river. Yet they wriggled away, heading directly for water even though a rise in the ground had to be climbed in order to reach it.

Eventually, one autumn, the time comes for them to spawn. The males, now about fifty centimetres long, may have spent only three years in the rivers; the females in the higher reaches may have been there for as many as eight or nine and are three times as long. But both are now heavy with fat. Once more their bodies begin to change. They turn from yellow to black. Their eyes begin to enlarge, suggesting that soon they will have particular need of them. They start to descend the rivers, resting on the bottom during the day and travelling mostly at night. The urge to return to the sea is now so great that they will wriggle out of a pond and cross dew-drenched meadows if that is necessary to reach a stream that will lead them to salt water.

Exactly what they do when at last they reach the sea was unknown until biologists charted some of their movements by persuading them to swallow tiny radio transmitters or inserting such devices beneath their skin. This research showed that they swim away from the European coast in a roughly north-westerly direction at a depth of about sixty metres until once again they reach the edge of the continental shelf. There, where the sea floor suddenly drops to 900 metres and more, they dive down to about 400 metres and swim off in a southwest direction.

This course certainly points them towards the Sargasso Sea and that indeed is where they are going. It has proved surprisingly difficult, however, to demonstrate this, because of the difficulty of capturing adult eels in the mid-Atlantic. This, presumably, is because they make the journey at very great depths, far below the reach of drift nets or trawls and since they are no longer feeding there is little chance of catching them on baited hooks or in traps. Eventually, some six months later, they reappear in the Sargasso. There they spawn. That done, they die.

Why should they make this journey of six and a half thousand kilometres, just to lay their eggs and by doing so commit their young to make a similar journey in reverse to find their feeding grounds? The answer, like the explanation of bird migration, lies in the past, though a much more distant one, for fish are a much more ancient group than the birds. Fossils of eels have been found in rocks a hundred million years old. These lived in the sea as many of their modern relatives still do. At that time, the continents of North America and Europe were still close together and the Atlantic was no more than a narrow strip of sea between the two. Some eels may then have discovered the richness of food to be found in estuaries and rivers and started spending the main part of their lives there. Since then, the forces of continental drift have continued to pull the two continents apart, widening the Atlantic, but the habit of returning to the sea each year to spawn has never been broken even though it now involves such an immense journey.

And what guides the eels on these amazing marathons? The young cannot learn how to make their first long journey by following their parents as swans do, for no adult eels travel east across the Atlantic from the Sargasso. Nor can the elvers have any memory of the smell of freshwater rivers, as salmon have, for none of them have been in one. Perhaps they are carried eastwards by the Gulf Stream in spite of anything they themselves may do. Maybe they even try to swim against it which might account for the length of time that passes before they reach European waters. Thereafter, a physiological change may cause them to develop a preference for fresher water so they are lured up the rivers, just as spiny lobsters, at a particular time of the year, are drawn to lower temperatures.

And the return journey? Experiments with captive eels suggest that they can navigate by the stars, just as some migrating birds can, and on the first stages of their journey away from Europe, when they swim near the surface, this may be the way they guide themselves. But what happens at the edge of the continental shelf? How, to start with, do they know that they have reached part of the sea where the water below them has suddenly become very much deeper? One researcher has suggested that they may be able to detect the very low frequency vibrations that are created in the water by waves and reflected back from the ocean floor, an echo-location system similar in principle to that used, at the other extreme end of the frequency scale, by bats. As they dive, so they once more encounter the Gulf Stream. But it is unlikely that they are able to use this in itself as a guide. No animal can know that it is in a current unless there is some stationary object, such as a river bank, to serve as a reference point. Eels are not thought to go down to the abyssal depths of the Atlantic where they could get such indications from the sea floor. Instead, recent research has shown that they get their reference points from the earth's magnetic field, just as pigeons do.

Finally, as they near the Sargasso, the imprinted memory of the particular smell of those strange waters may be sufficient to lure them to the very same semi-stagnant tract of ocean where they hatched, just as adult trout are led to their own birth-places.

Although the mysterious eel has begun to reveal some of its secrets, we still have a great deal to learn about the skills that animals use to find their way around their own home ranges and to travel the globe.

SIX

Home-making

ew places have such a gentle and equable climate that animals living there never need seek shelter; and few animals are so well-armed that they do not welcome somewhere to hide from their enemies or a safe nursery for their young. So many animals, at some time in their lives if not throughout them, need a home. To create one, they become potters and plasterers, weavers and needle-workers, miners, masons, scaffolders, thatchers and sculptors. These crafts are not necessarily the monopoly of any one group of animals. Each species, within the limitations of its own anatomy and the physical possibilities of its surroundings, uses the technique most appropriate to its needs.

The simplest home of all, of course, is no more than a hole. A branch falls from a tree, letting fungal decay into the heart of the trunk, and there is a home for owls and flying squirrels, lemurs, parrots and toucans. A river with slightly acidic water dissolves its way into limestone, and, once the waters recede, there are even larger holes – caves – for bats and bears, even in some places and at some times, human beings. But naturally occurring holes are in relatively short supply. Most hole-dwellers have to excavate their homes for themselves and that can be hard work.

The piddock drills its hole into solid rock. It is a small mollusc, the size of a mussel, that starts life as a tiny free-swimming speck of jelly. This larva eventually settles down on a rock, usually chalk or limestone, and grows the two valves of its shell. Chemically, these are largely calcium carbonate, but at one end the edges are armed with small hard spikes like the teeth of a saw. The young piddock clasps the rock face with a muscular sucker called a foot, presses its saw teeth against the rock and starts to swivel back and forth. The teeth bite into the stone as the animal, slowly, methodically and persistently, continues to bore inwards. Within a few days, it has created a shaft so deep that it is out of sight and beyond attack. From this secure position, it extends a long tube, its siphon, along the tunnel and into the open water to suck in a current that brings with it minute particles of food.

Improbable though it may seem, birds also manage to dig into stone. They have only one tool with which to do so, their beak, but this can be surprisingly effective. The bee-eater's beak is slender and apparently delicate, a pair of slim forceps with which it plucks bees and other insects from the air. Yet when the bee-eater starts to make its nest, it will fly repeatedly beak-first at the face of a sandstone cliff or the hard mud of a river bank until, by dislodging grain after grain, it manages to make a slight depression to which it can hang. Thereafter it hammers away until it has excavated a narrow tunnel as much as a metre long.

Several species of bee-eater nest in colonies a thousand or more strong. This may be because of a scarcity of suitable sites, but the great number of individuals present makes it possible for unmated young birds to help their parents in the labour of digging holes just as the young Florida scrub jays help their parents rear new broods. Artfully and sensibly, the red-throated bee-eater of Nigeria starts its labours at the end of the rainy season when the ground is relatively soft, even though it will not be ready to lay its eggs for another three months. Woodpeckers, accustomed to chiselling their food out of timber, have little difficulty in cutting out nest chambers in tree trunks. They make such excellent roomy holes that often other animals, lacking the tools for carpentry but otherwise of a powerful disposition, such as owls and squirrels, will drive away the woodpeckers and claim the holes for themselves.

Reptiles also make tunnels. The gopher tortoise that lives in the southwestern deserts of the United States needs one as a shelter in which to escape the worst of the mid-day heat and it digs into the sun-baked ground with slow ponderous sweeps of its armoured fore-legs. These tortoise holes are often so long – up to twelve metres – that judging from the tortoise's slow rate of excavation they must have been made by several generations and are probably several centuries old.

Small mammals too are great diggers. The North American kangaroo rat and the South African springhare (a large rodent that is not a hare) use their holes, as the tortoise does, to shelter from the heat; hyaenas and wolves as nurseries; badgers and armadillos as dormitories in which to slumber during the day after foraging at night; and mice and rabbits as sanctuaries where they are beyond the reach of most of their enemies.

But holes with only one entrance can also be death-traps. An animal sheltering in them can only too easily be cornered and many hole-dwellers take steps to reduce that risk. Nuthatches, nesting in tree holes, habitually narrow the entrance by rimming it with mud so that nothing bigger than they can enter, and even the probing beak of a magpie hunting for nestlings cannot reach the young within. Most hornbills go even further. When the female settles on her eggs, the male brings lumps of soil, moistened with his saliva, as she sits in the hole. With this the two of them wall up the entrance until only a tiny slot is left open. For the next few weeks the male passes food for his entire family through this until the female breaks the wall down and leaves the nest-hole to help him collect the increasing quantity of food demanded by the growing chicks. Once their mother has left, the chicks themselves rebuild the wall. Only when they are fully fledged and ready to fly do they demolish it and leave.

Perhaps the best protected hole of all is that built by the female trapdoor spider. In some species she is about two and a half centimetres long and she digs a burrow fifteen centimetres deep into soft ground. She uses her silk to line the walls and also to bind particles of soil together into a circular lid two centimetres across. She gives this a silken hinge and attaches gravel to its underside so that it will fall shut under its own weight. Since it is made from local material, it exactly matches its surroundings and it fits so neatly with a bevelled edge that it is almost impossible to detect it. Nor will the spider give away the presence of her home during daylight hours. Only when evening comes does she lift the lid a chink and peer out, checking whether darkness has yet come. When night does fall, she opens the lid and stretches out her two front pairs of legs. If an insect walks by, she grabs it and instantly drags it inside the tunnel. The weighted door slams shut and she consumes her prey in safety.

So secure is this home that once the female spider has built it, she never leaves it. She even, on occasion, locks herself inside it. As she grows, she must periodically shed her inelastic skin. Immediately after doing so, before her new skin has had a chance to harden, she is particularly vulnerable, so before the event, she ties down the door from the inside with ropes of silk.

The males build similar holes, but leave them to visit the females and mate with them in their tunnels. After that is over, the male slips out of the door and the female once again locks it with silk. Then, confident that she will not be disturbed, she retreats to the bottom of the tunnel to lay her eggs.

Holes, particularly long ones, have another drawback. They can get very stuffy. Prairie dogs, rabbit-sized rodents with short legs and small ears that live in vast communities on the grasslands of the American West, dig tunnels that may be as much as twenty-seven metres long with short cul-de-sacs on either side. Each tunnel has two openings, one at each end. That, in itself, helps. But the arrangement is more artful than it might appear to be at first sight. The two entrances are not the same shape. One opens flat on the surface of the prairie. The other is enclosed by a chimney of mud and stones that may stand as much as thirty centimetres tall. Wind moves faster a little above the ground than it does close to the surface, so a breeze blowing across the top of the chimney sucks out the stale air within the tunnel and draws in fresh air through the lower entrance.

You can easily demonstrate the efficiency of this system by lighting a harmless smoke candle close to the lower entrance. The smoke is drawn gently into the hole and then, several minutes later, wisps of it emerge from the top of a chimney fifteen metres away.

Digging holes, even holes with a relatively sophisticated design such as the prairie dog's, requires strength rather than ingenuity. Building a home is a more demanding business altogether since it necessarily involves assembling the right materials, deciding on a layout and somehow fastening things together. Mammals, by and large, do not tackle these problems. Holes, it seems, give them all they need. One of the few exceptions is the beaver.

Beavers live in the forests of North America and in many parts of Europe and have recently been reintroduced into the United Kingdom. They feed on leaves and the living bark of trees. To get what they need, they cut down saplings and even trees with trunks thirty centimetres or so in diameter, gnawing through the wood with their chisel teeth. They need a home where they will be safe from hunting animals such as bears or lynx and they also need a place to store food for the winter when the land is snow-bound. They get both by building a dam.

A newly-mated pair of beavers choose for their home a valley with a small stream running down it. With a fine eye for the lie of the land, they select a particular point along the stream and then start to build their dam. They ram sticks upright into the stream bed. Across them, they drag thin saplings and then trundle boulders on to them to weight them down. They dig mud from the banks and nudge that into the construction to bind the sticks, leaves and boulders together. If the supply of suitable saplings near the site runs out, they will dig canals leading into the stream and float down their logs from farther away. As the dam grows, they give its sides a slightly different character. The up-stream side becomes steep and heavily plastered with mud to make it water-tight. The down-stream side is more sloping and covered with poles laid parallel to the sides of the valley, so giving the structure the strength to withstand the pressure from the water that accumulates in the lake. And at each end of the dam, they clear a spillway.

On the shores of the lake, or on one of the little islets that may form in the middle of it, they build their lodge, a great dome of sticks, poles, branches, reeds and mud within which they have their living chamber. It is the dam, however, that makes this virtually impregnable for the only way into it is from the lake through a tunnel that opens underwater. So only agile swimmers like beavers can enter it.

Securing the lodge is not the only function of the lake. As autumn approaches, but while the trees are still in leaf, the beavers cut down saplings and sink them in the lake. There they lie, hardly decaying in the near-freezing water. When snow metres deep blankets the land and the lake is covered with ice, the beavers are able to swim out from their lodge beneath the ice, retrieve the green branches and feed on them throughout the winter.

The maintenance of the dam requires the constant attention of the owners. If there is heavy rain, the spillways must be enlarged to allow the flood water to escape before the dam bursts. And when the rain stops, the beavers may have to build them up again to prevent the level of the lake from falling so low that the entrance to the lodge is exposed. Many of these lakes last for decades if not centuries and are used by several generations of beavers. Ultimately, however, the beavers' lake, like any other, will fill with sediment and turn first into a swamp and then into a level green grassland. No doubt when people moved into these forests for the first time, they were delighted and surprised to find such fertile meadows in the heart of the dense forest and built their homes beside them. So the predilections of beavers centuries ago may well have determined the places where human beings have their towns today.

Placing sticks on one another in such a way that they engage and are not readily dislodged is a skill that is easy to underestimate. You realise how subtle it is when you watch a beaver painstakingly placing a pole on its dam, being visibly dissatisfied with its position, removing it and tugging it into another, until it is at last convinced that it has been effectively deployed. Birds possess a similar skill, as you will discover if you try to dismantle even the untidiest and most apparently haphazard of twig nests. Most of the elements interlock with one another. Often there is a symmetry underlying the superficial irregularity of the twigs, a basic radial pattern or an equally deliberate criss-cross arrangement.

Although large tree-nesting birds – wood pigeons, rooks, storks or eagles – do only the minimum to soften the uneven surface of their platform nests, many smaller birds with delicate eggs fashion a cup in the centre of the nest which they line with softer material. Most species have their own particular taste in this. Thrushes use mud, the bearded tit likes flower petals. The Australian honeyeater is so fond of hair it will pluck it straight from a horse's back or even a human's head. North American house wrens favour sloughed snake skins and the eider duck grows special downy feathers on its own breast which it plucks off with its beak to produce a soft and warming blanket that none of our synthetic materials can equal.

Some birds are so small that twigs are too coarse a material for them to use as the main fabric of their nest. Hummingbirds use spiders' silk, gathering it in their beak and flying away with it trailing behind them. Their ability to hover enables them to construct their tiny cups on sites that are not even big enough for them to perch on – two crossing stems, perhaps, or even the tip of a leaf. Exploiting the glueyness of spider silk, a hummingbird will prod it repeatedly at a selected spot until at last it sticks. If the nest is a hanging one, then the bird will fly round and round the first threads, binding more around them to form the walls of the cup. Often they add flower petals or tiny fragments of down or lichen to give the construction body. If the nest is attached by one side to a leaf, its support is clearly lop-sided, in which case the hummingbird may weave small particles of earth into a long extension of the nest, dangling beneath, to act as counterweights and level it. This gives the little nest additional stability and so reduces the risk of it being overturned by a gust of wind.

The Indian tailor bird also uses spiders' silk but in a different way. It sews with it. It creates a cup from living leaves, either from two hanging closely together or by twisting a single one into a curl. Holding a length of silk in its beak, it then pierces a hole in the leaf and pushes the silk through it, tying a little knot in the end to prevent the thread from slipping back through the hole and then doing the same thing on the other side so that the two leaf surfaces are secured to one another.

To describe this as sewing is perhaps slightly flattering to the bird, for the silken thread is never used for more than one stitch at a time. Other birds, however, can be said with accuracy to weave, for they create the fabric of their nests using exactly the same principles as human weavers employ when they interlace a weft thread between parallel warp threads to create cloth. This technique has been discovered independently by two different groups of birds – the weaverbirds of Africa which are closely related to the European sparrow, and the icterids of the Americas which include such birds as the caciques and the oropendolas. The fibres they use may be long creepers, thin rootlets, ribbon-like leaves such as those of grasses or reeds, or strips torn from broad leaves such as bananas.

Two basic skills are needed – knotting and weaving. A knot is required in order to make the first fastening. The bird ties it by holding a strip on to a branch with one foot and then, using its beak, passing the end round the branch, threading it through one of the turns and pulling it tight. Sometimes the bird may wind the strip around not one but two parallel twigs. Then the knot is made even more secure by threading the end between the two twigs and tying a series of half-hitches on each.

The fastening finished, the bird starts weaving. The process involves no more than threading a strip beneath another one that runs across it more or less at right angles, and keeping on doing so at intervals with dogged persistence, pulling the strip tight after each threading. Sometimes if the strip is long enough, the bird will reverse direction during this process so that the strip is looped back and woven parallel to itself. This produces a particularly firm fabric.

With these two basic skills, bird weavers can attach nests that dangle from tips of branches or leaves and construct domed and compartmented dwellings of great perfection. Some nests are given waterproof roofs by using particularly wide strips of leaves for the top half. An antechamber may be built on to the main egg-chamber. Some species also add entrance passages, long downward pointing tubes that make it extremely difficult for snakes or any other intruders to plunder the nest. One of the most skilful craftsmen of all, Cassin's weaverbird, constructs an entrance-pipe sixty centimetres long using very long narrow fibres that spiral downwards, some right-handedly and some left-handedly, so that the two interweave with one another and produce a fabric of remarkable uniformity and beauty.

Weaverbirds reared in an incubator still manage to weave when they become adult, so clearly the basic skill is inherited, but it nonetheless requires practice to bring it to perfection and at first young male weavers may make comically ham-fisted versions – nests that are insecurely tied and fall off, others that are unevenly woven with some strips pulled tight and others left slack so that the result is misshapen. The males' skill affects their breeding success, for the females examine these nests extremely critically when selecting a mate. The hopeful male hangs beneath his creation fluttering his wings in order to draw attention to it. If it is not of reasonable competence, no female will join him. In such circumstances, he has to start all over again. Since both sites and nesting material are usually in short supply in a weaverbird colony, he often laboriously unravels his first attempt and using its constituents weaves it all over again in the same place.

One group of birds has particular problems in nest-making. Swifts are the most aerial of birds. They spend months on end continuously on the wing, feeding by catching insects in mid-air, mating by coupling high in the sky and tumbling downwards interlocked for dozens of metres, and presumably even sleeping on the wing. But they cannot incubate their eggs in the air. To do that they have to come to earth and sit on something solid. That is not easy for them to do, for so extremely have their bodies become adapted to life in the air that their legs are greatly reduced in size. They are little more than delicate hooks hidden in their plumage and are so short that they are quite incapable of lifting the bird's body high enough to enable it to make a complete wing beat. So swifts, if they land, find great difficulty in getting back into the air and when the time comes for them to make a nest, they are unable to gather leaves, sticks or mud in the way that other birds can.

The chimney swift of Asia manages to collect twigs by flying at a thin branch, seizing it with its beak and breaking it off by the sheer force of its aerial velocity. It then glues these thin twigs to a wall using its own saliva as a fixative. The American palm swift also produces a sticky saliva but makes no attempt to gather anything as substantial as twigs. It builds its nest entirely from air-borne material such as cotton, plant fibres, hairs and feathers. The African palm swift hardly bothers even with these and constructs its nest almost entirely from its saliva, moulding it into a tiny spoon-shaped structure, stuck to the underside of a palm frond.

When the palm leaf sways in the wind, it seems almost impossible that the single egg could remain in the tiny cup. Indeed, it would certainly fall out were it not for the fact that the bird has not only glued the nest to the leaf, but the egg to the nest. The nest is far too small for a bird to sit in. Instead, the incubating parent has to stand astride it. Nor can the nest accommodate the chick when it hatches. The young bird has to perch upright on the rim while it grows its feathers.

The echo-locating cave swiftlets of South-east Asia also build with saliva, but they are more lavish in their use of it. They have very well-developed glands in their throat that, during the breeding season, become greatly enlarged and produce saliva in considerable quantity. Although the several species that nest near the mouths of caves incorporate feathers into their constructions, the one that inhabits the deepest and totally lightless parts of caves constructs its nest from saliva and nothing else whatever. It will build this on a ledge if one is conveniently available, but it is perfectly capable of fixing the nest to a vertical or even an overhanging wall of rock.

The bird starts by flying persistently in front of its chosen site and repeatedly dabbing the rock with its tongue, laying down a curved line of saliva which marks the lower edge of the nest-to-be. The saliva dries and hardens quickly and with repeated flights, the bird slowly builds up the line into a low wall. As soon as this is big enough to cling to, the speed of construction accelerates and within a few days the wall has become a semicircular cup of creamy white interlacing strings that is just big enough to hold the customary clutch of two eggs.

Producing material from glands in the body is exceptional among birds, but among insects it is almost the rule. Indeed the silk that we ourselves spin and weave into the most luxurious of all our fabrics is unwound from the cocoon that silk moth caterpillars weave around themselves before they start the complex process of changing into adults. They produce this silk not from spinnerets at the end of the abdomen as spiders do, but from a pair of glands in the mouth. Producing it makes considerable demands on an animal's bodily resources and although the silk moth, encouraged by our selective breeding, extrudes it in large quantities, most insects are rather more sparing in their use of it. Ermine moths, for example, economise by constructing a cocoon that is little more than a lattice.

Ants, like moths, only produce silk during their larval stage, but this does not prevent them from using it for nest-building. The green tree ants of Australia construct their homes from the living leaves of trees. Squads of workers hold two leaves together, gripping their rims with their legs, while others dash to the nursery quarters. There they pick up the little grubs in their jaws and carry them back to the building site. A worker stimulates the grub to produce its silk by giving it a little squeeze.

Then it passes this living tube of glue back and forth across the junction of the two leaves until a white silken sheet has been created, linking them.

Bees too secrete their building material. Flakes of a fatty substance are excreted from glands between the joints on the underside of the worker bee's abdomen. It collects these flakes with the brush near the end of its hind legs and passes them forward to its mouth where it kneads them with saliva. This material is now wax and with it the worker bees build combs that serve both as nurseries for the developing young and larders for honey and pollen.

A honey bee colony, which may eventually contain up to eighty thousand individuals, is founded when a young queen hatches in an existing colony and emigrates, taking half of the workers with her. They find a site for a new nest, such as a hollow tree (or a beekeeper may provide them with one), and start immediately to build combs. For hundreds of years, scientists and naturalists have been struck by the fact that the cells which make up the comb are generally uniform in both design and dimension. They are six-sided with walls that meet at 120°, with the base of the cell formed by three rhombuses. This design cannot be dismissed as an automatic and inevitable consequence of building cells closely together, for bumble bees also make waxen cells and theirs are irregular bag-shaped containers, jumbled untidily together. The honey bee, in contrast, has developed a special, highly sophisticated skill by which it uses both the wax and the space within its nest in the most economical way possible.

Were the cells circular, there would inevitably be gaps between the walls of one cell and the next. No matter how you pack snooker balls on a table, there will always be space between them. The only shapes that fit together so closely that all their walls are common with those of their neighbours are triangles, squares and hexagons. Of these three, the hexagon has the shortest total length of wall for a given enclosed area. Furthermore, the shape of the rhombuses at the base of the cell maximises the storage capacity. Although each cell does not begin as a precise hexagon, but rather as a form of cylinder, it gradually takes on a hexagonal shape as the comb is built. This shape saves building materials.

Nor do the bees make the cell walls thicker than is necessary to carry the stresses they have to bear. The combs hang vertically with the hexagonal cells on them facing outwards and tilted slightly upwards from the horizontal so that the honey, before it is capped off, does not run out. The bees work, not on a single cell at a time, but on whole sections of the comb. Each cell wall is first laid down as a thick ridge. The worker then reduces it to its proper thinness by putting her head in the cells on either side of the wall and shaving off layers from the face of it. She then measures the thickness of the wall by pressing it with her mandibles and detecting how much it bends. Since the temperature within the nest is kept constant and the composition of the wax is uniform, the degree to which the wall bends under the same force is an accurate measure of its thickness.

A bees' nest has only a single entrance, so air cannot be made to flow through it as it does through a prairie dog's tunnel. Instead, the colony has to inhale and exhale. When the level of carbon dioxide breathed out by the bees gets high in some part of the nest, groups of several hundred workers start vigorously fanning with their wings, so circulating the air around the combs and levelling out any imbalances. If the temperature also rises above the bees' favoured level of 35°C, a group of workers position themselves near the entrance with their tails pointing outwards and then start fanning their wings so that the stale air is wafted out. After doing this for perhaps ten seconds, the fanners beside the entrance stop simultaneously and fresh air is drawn back into the hive. In this way the nest can, as it were, take three breaths a minute.

The bees also have other ways of maintaining an even temperature within the nest. If it rises too high, the workers bring in not nectar but water, and deposit it in droplets and puddles around the cells containing the developing larvae, which are particularly sensitive to over-heating. They then fan the water so that it evaporates and in doing so lowers the temperature. If on the other hand, the colony gets too cold, as it may in winter, the workers eat honey and use its energy to vibrate their flight muscles within their thorax without moving their wings, so generating body heat.

Wasps also make nests filled with combs of hexagonal cells, but they build not with wax but with paper. This they make by chewing wood, masticating it with their saliva and then expelling it as a moist pulp which hardens as it dries. The resulting parchment is both very strong and very light and with it, paper wasps can build very big nests indeed. The combs do not hang vertically like those of honey bees and they contain neither honey nor pollen, for wasps are not vegetarians but primarily carnivores. The cells are filled only by developing young which the workers feed with pellets of chewed-up caterpillars or other bits of flesh.

A queen wasp herself selects the site for her colony. Tropical species often use open sites, under a branch or even a large leaf. The common European wasp favours a cavity – a tunnel dug by a field mouse or some other small mammal, a hole in a tree, or a warm place in the loft of a house – and there she fixes a stalk of paper to a point in the roof. On the lower end of this, she constructs a small group of downward-facing cells in each of which she lays an egg. These, when they hatch, provide her with her first work force. Soon these young wasps are also busy, chewing wood, producing paper and building cells in which the queen lays more eggs.

Whereas the vertical combs of honey bees have cells on both sides, the paper wasp's horizontal combs have them only on the under-surface. The way the wasp builders calibrate their work is also different. Instead of planing the sides of the cell-wall under construction, the wasp extrudes the paper pulp along the line of the wall and then, working along the rim with her mandibles, she judges the angles and the thickness of the wall by continually touching the walls of the adjoining cells with her antennae.

As each comb is completed, rods of paper are fitted to it, extending vertically downwards, from which the next horizontal comb will be suspended. If the nest is underground some of the workers gnaw at the soil of the floor, carrying it away cradled in their forelegs and dropping it outside. If they come across pebbles too big to lift, they dig away beneath them so that the pebbles slowly sink, millimetre by millimetre. When the great globular nest is complete there is often, therefore, a layer of gravel and little pebbles at the bottom of the chamber.

Other wasps build with mud. These are the potters, the females of which make small jars in which they deposit an egg together with a paralysed spider or caterpillar to provide the hatching grub with its first meal of meat. A female starts by gathering mud from a patch of wet ground. As every potter knows, the moistness of the clay is very important and the potter wasp controls it very precisely. If it is too dry, she regurgitates water from her stomach and moistens it. She then kneads it with her mandibles and front legs into a soft pellet about half the size of her head and flies off with it to her building site. This may be under a piece of tree bark, along a ledge of some kind or hidden on the ground among the undergrowth. With scissoring movements of her legs and mandibles, she converts the pellet into a long glistening strip of clay and lays it carefully in a ring. She works fast, adding one strip to the rim of the hardening strip below until she has built a small bottle, complete with an elegant out-turned lip to its mouth. When she has stocked it with an egg and its immobilised meal, she seals the vessel with a bung made from one final pellet.

Birds also pot. Swallows mix their mud with grass so adding to its strength, and pile pellet on pellet to build cup-shaped nests on ledges beneath house eaves. These are so strong that, since they are usually sheltered from the rain, they can be used, after judicious refurbishment, year after year. The rufous ovenbird, a South American species rather like a thrush in shape and size, builds on a very large scale indeed. Its completed nest is the size of a football, domed with a slot-shaped entrance on one side. Put your fingers inside and you will feel, not eggs, but a dividing wall. The only way past this into the egg-chamber beyond is through a small hole high in one corner and neither your fingers nor an intruding paw or beak can negotiate the bends and reach through it.

But the most skilled and ingenious of all mud-builders are termites. There are over two thousand different species of these insects, virtually all of them feeding only on dead plant material. Most have bodies with a skin so soft, thin and permeable that if they are exposed to direct sunshine for any length of time they dry out and die. So they spend their lives in darkness, and, having no use for eyes, are totally blind. Some chew galleries inside trees or house timbers and digest the wood they excavate with the help of micro-organisms in their gut. Others burrow away underground to retrieve fragments of dead plants from the soil. Some make combs of their waste material on which they cultivate tiny fungal mushrooms. Many can forage above ground at night but they laboriously shield themselves from predators by constructing covered run-ways, and thin crusts of mud all over the vegetation they are plundering. And some masticate earth with saliva to produce a cement that sets as hard as rock and with it build some of the most magnificent and complex mansions to be found anywhere on earth except for those made by human beings.

The foundations of these buildings are laid down when a mated pair of king and queen termites crawl into a crevice in the ground, burrow downwards and start laying eggs. The queen grows in size, and thereafter for the rest of her life continues producing eggs which hatch into workers and soldiers. It is the workers who build the colony's home, burrowing out galleries deep in the earth and, above ground, raising great domes, towers, turrets and spires.

The air-conditioning of these buildings is of crucial importance to their inhabitants. The communication between the members of the colony depends on a system of chemical exchange that is badly disrupted if the temperature rises too high. If the air becomes too humid, fungus may germinate in their stores of dried vegetable food and ruin the crop. Most crucial of all, the royal couple themselves, with their thin permeable skins, will die if the atmosphere gets so hot and dry that they lose their body fluids or if they are badly chilled, and this may mean the end of the whole colony.

But it is very difficult, if not impossible, to make the interior of these huge homes totally impervious to extreme changes in the weather outside. The inhabitants, therefore, may have to move from one part of the nest to another to find the most comfortable conditions. If the night is very cold or the middle of the day

extremely hot, they may retreat to the galleries below ground
level where the temperature does not vary so much.

In parts of northern Australia, where there are heavy seasonal
rains, the ground becomes so waterlogged for part of the year
that these downward migrations are not possible. One species,
however, has evolved a solution. A colony builds a hill in the
shape of a rectangular wedge standing four metres or so high
with its sharp edge at the top. Each side may be three metres
across and the hill is aligned so that the thin edge of the crest
points north and south. This species, therefore, is known as
the magnetic termite. But the stimulus that leads the termites
to orientate their building in this way has nothing to do with
magnetism and everything to do with heat.

In the morning the colony may be very cold, for at night the
temperature here can fall to within five degrees of freezing. But
one broad flank of the colony faces east and so catches the full
warmth of the rising sun. At this time the termites congregate
in their eastern galleries. As the sun rises and the day warms
up, the outer surface of the hill may become so hot that it is
almost painful to touch, but the full strength of the midday sun
is minimised since all it strikes directly is the thin top edge of
the hill. As the sun sinks and the day cools, so the western flank
is illuminated, the east is in shadow and the termites within can
find just those galleries between the two that best suit them.

In sober truth, although all the magnetic termite hills in any
one area stand parallel to one another, they do not always point
accurately north and south. There can be a variation of up to
ten degrees either east or west. The explanation of this is that

the sun is not the only agent that affects the temperature of the colonies. The prevailing wind, the shape of nearby hills and many other things also have an effect and make it more advantageous thermally for the hills to point slightly to one side or the other of magnetic north. Since the termites are responding not to magnetism but to heat, they build accordingly.

This is not to say that these termites have no perception of the earth's magnetic field. On the contrary, they and indeed most other termites certainly have. Experiments have been made in which powerful magnets were placed around a nest to distort the magnetic environment. The termites continued to build their hills with their crest pointing in exactly the same direction as before, but they did change the disposition of their elongated chambers within the nest. It seems that the construction workers building galleries in total darkness get their bearings and coordinate their work by sensing the magnetic field of the earth.

Those termites that live in parts of the tropics where it is wet throughout the year have different problems. In such places the abundant rains produce a tall thick forest beneath which the air is constantly warm and humid. They do not therefore have to deal with great daily variations in these conditions. Their main danger is that their entire home will become drenched and waterlogged. So here termites construct circular towers with conical roofs, the eaves of which project outwards so that the rain cascading from the tower falls some way from its base. Each additional storey may be given its own roof so that eventually the building looks like a Chinese pagoda.

The bigger the mound, the greater the need for air-conditioning and one of the biggest of all is built by an African species with particularly aggressive soldiers, known as the bellicose termite. But this particular species has an additional need to control the temperature within its habitations. Many other termites digest their unpromising food of twigs and vegetable detritus with the aid of micro-organisms in their gut. The bellicose termite belongs to a group that uses a different digestive method. They employ a fungus to do it for them. The workers eat almost nothing but dead wood. Although they absorb a little from their first meal, their droppings still contain a great deal of unextracted sustenance. So they defecate in special chambers within the nest and on their manure cultivate their fungus. Its hair-thin filaments permeate the manure, absorbing much of it, changing the nature of the residue, occasionally producing tiny reproductive organs like white pin-heads. After the dung has been processed in this way for about six weeks, the termites are able to eat and digest it, fungal threads, white knobs, residue and all.

The fungus that performs this service belongs to a group that lives nowhere except within the nest of termites and each species of fungus-grower cultivates its own unique species. The termites are totally dependent upon it, just as it is dependent upon the termites. And it grows best within a very precise temperature range of 30–31°C. However, the processes of decay in the gardens produce a great deal of heat. So do the million and a half termites living in the colony. They also impoverish the air by inhaling oxygen and exhaling carbon dioxide, as all animals do. For a colony of fungus-growing termites efficient air-conditioning is therefore vital.

Their way of providing it is architectural. Although the climate is much the same throughout the range of the bellicose termite, the soils are more variable and the species adapts the design of the nests to suit the position and strength of the local building material. Some are massive domes two metres high, some are low

Mountainous star coral, spawning at night, releasing vast numbers of both eggs and sperm, Grand Cayman.

Above: *Female land crabs descending to the sea to spawn, Christmas Island, Indian Ocean.*

Below: *A tarantula hawk spider wasp, having stunned a tarantula, hauls its victim back to her burrow, California.*

Above: *A killdeer plover shading eggs from the sun on her nest, Welder Wildlife Refuge, Texas.*

Below: *Male mallee fowl digging to regulate the temperature of incubating eggs, Victoria, Australia.*

A female aphid giving birth, Pennsylvania.

Above: *A male Korean seahorse releasing his babies, Kyushu, Japan.*

Below: *An adult reed warbler feeding a cuckoo chick that has ousted the warbler's chicks and now occupies the whole of the nest, the Fens, Norfolk, England.*

Above: *A female opossum carrying the survivors of her large brood, Tortuguero National Park, Costa Rica.*

Below: *The arrival of the wildebeest on the savannah of Masai Mara, Kenya.*

Above: *An elephant seal bull, female and pup, St Andrew's Bay, South Georgia.*

Below: *Mara group with young, Valdes Peninsula, Patagonia.*

Above: *Mallard ducklings following their mother, Northern Ostrobothnia, Finland.*

Left: *Mexican free-tailed bats emerging from Bracken Cave at dusk, Texas.*

Right: *A lion cub tackles a wildebeest caught for him by his mother, Tanzania.*

Below: *Emperor angelfish, Red Sea.*

Above: *A great grey owl with its differently-sized chicks, Scandinavia.*

Below: *African elephant calf and juveniles playing together, Masai Mara, Kenya.*

Above: *Garden bumblebee, flying to a foxglove flower, Wales.*

Below: *Replete honeypot ants, fully-laden, hanging in their underground gallery, northern Australia.*

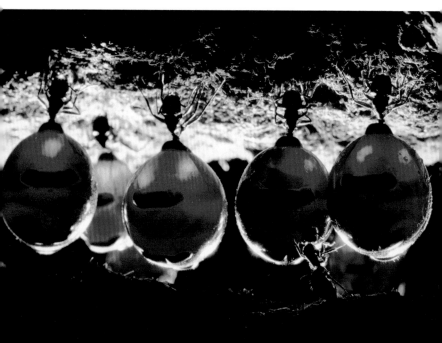

A leaf-nosed bat, sipping nectar from a banana flower, Costa Rica.

Above: *A green-crowned brilliant hummingbird feeding from a Heliconia flower, Costa Rica.*

Below: *An egg-eating snake, with its jaw unhinged, prepares to swallow an egg, South Africa.*

Red-and-green macaws on a salt-lick, Peru.

Above: *A female magnificent spider, swinging her trapline as a moth approaches, lured by the spider's pheromones, New South Wales, Australia.*

Below: *A gladiator spider, holding a net of silk between its four front legs, hangs in wait to cast it over a passing victim, Queensland, Australia.*

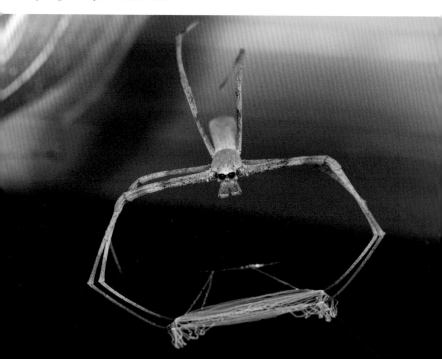

Above: *A killer whale attacking sea-lion pups, Patagonia.*

Below: *Army ants attack a cricket, Peru.*

Right: *Barracuda and bluefish circling a bait-ball of Atlantic mackerel, Azores.*

Below: *A fire-bellied toad giving warning of its poison, East Asia.*

Above: *A giant pronghorn hover fly, which mimics both the appearance and the behaviour of a wasp, England.*

Left: *A striped skunk has conspicuous black and white colouration to warn of the powerfully smelling liquid that it can squirt at those which might interfere with it, North America.*

Above: (left) *A cluster of tree hopper larvae that mimic withered flowers, Madagascar;* (right) *a swarm of adult tree hoppers that resemble thorns, Costa Rica.*

Below: *A grasshopper camouflaged to match the lichen on which it habitually sits, Costa Rica.*

A moth caterpillar that, when threatened, expands its hind end to mimic the head of a viper, Costa Rica.

Above: *A hairstreak butterfly which, when perched, habitually flickers elongations of its hind wings. These resemble antennae and so deceive a predatory bird into attacking its less vulnerable end, Sarawak, Borneo.*

Below: *A tree hopper with the colour and texture of the leaf which it habitually eats, Guyana.*

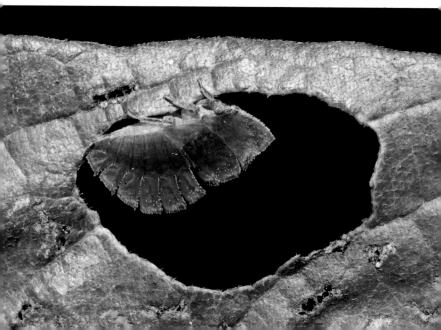

Above: *An angler fish near the surface of the sea among floating sargassum seaweed, West Papua.*

Below: *Chimpanzees feeding on a dead bushbuck, Tanzania.*

Above: *A hyaena pastes while another tastes and smells the message, Kenya.*

Below: *White-chinned petrel nesting, Kidney Island, Falkland Islands.*

Above: *A Daubenton's bat flying low over water, hunting, UK.*

Below: *Amazon river dolphins, one with a newly-caught fish in its jaws, in the flooded forest around the Rio Negro, Brazil.*

Above: *A wels catfish with long sensory barbs on the front of its head with which it investigates its surroundings, Cher River, France.*

Below: *Spiny lobsters migrating from their juvenile to their adult habitat, the Bahamas.*

Elvers, juvenile European eels, climbing a rock wall during their migration upstream, Henllan Bridge, Wales.

Above: *Carmine bee-eaters at their nest in a river bank, Namibia.*

Right: *Great Indian hornbill female, inside her nest, passing food brought by her mate, to their chick, Borneo.*

Above: *A giant trapdoor spider at the mouth of her underground silk-lined nest, Malaysia.*

Below: *A beaver at work underwater, Minnesota.*

Above: *A female green hermit hummingbird, feeding her chicks in the nest she has built on a leaf tip, Costa Rica.*

Below: (left) *A male southern masked weaver bird constructs the foundation ring of his nest, Western Cape, South Africa;* (right) *a male vitelline masked weaver, hanging from his completed nest, Kenya.*

Above: (left) *European hornets on their nest, France;* (right) *a female potter wasp completing the entrance to her bottle-shaped nest, Israel.*

Below: *A worker ant building a nest and using the liquid silk produced by a larva to stick leaves together, Borneo.*

Above: *Magnetic termite hills, Northern Territory, Australia.*

Below: *A termite mound with cooling chimneys, Mboko, Congo.*

A male golden-shouldered parrot, perched on top of the termite mound in which he has his nest, Cape York Peninsula, Queensland, Australia.

Right: *Pompom crab with living anemones attached to its claws. It uses them in defence; and they benefit by being continually moved to new feeding grounds, Hawaii.*

Below: *Three false clownfish settled in a sea anemone, Cabilao Island, Philippines.*

Above: *Green tree ants milking honeydew from the caterpillar of a gossamer-winged butterfly, Australia.*

Below: *Moth butterfly larva feeding on weaver ant larvae, Queensland, Australia.*

Above: *A newly-born three-toed sloth clinging to its mother whose fur contains a permanent population of moths, Costa Rica.*

Below: *A tightly-packed roost of eastern long-fingered bats, Queensland, Australia.*

Above: *A Darwin's finch collecting ticks from a giant tortoise, Isabela Island, Galapagos.*

Left: *A moray eel opening its white mouth so that it may be cleaned by a cleaner wrasse, Christmas Island.*

Above: *Female harlequin beetle covered in parasitic mites, Trinidad.*

Below: *A garden snail with bands on its abnormally club-shaped tentacles, caused by a parasitic fluke living within its body. The strange appearance attracts the attention of birds such as thrushes, which eat the snail, transferring the fluke to a second host, Norfolk, England.*

Above: *Griffon vultures fighting over a zebra kill, Masai Mara, Kenya.*

Below: *Beadlet sea anemones touching tentacles, the Channel Islands.*

Opposite: *Zebra stallions fighting, Etosha National Park, Namibia.*

Right: *A male panther chameleon, in a defensive posture display, Amber Mountain National Park, Madagascar.*

Below: *Pair of female blue poison-dart frogs fighting, South America.*

Above: *Male giraffes necking to establish dominance, Masai Mara, Kenya.*

Below: *Speckled rattlesnakes fighting, California.*

Above: *Moose bulls sparring, Grand Teton National Park, Wyoming.*

Below: *Stag beetle males fighting on an oak tree branch, Elbe, Germany.*

Above: *Alpine ibex adult males fighting on a steep mountain side, Aosta Valley, Italy.*

Below: *A timber wolf alpha male asserts his dominance over another pack member as they squabble over the body of a white-tailed deer, Minnesota.*

Above: *An argument between whooper swan families, Martin Mere Reserve, Lancashire, England.*

Below: *A male great tit perched with spread wings in a show of aggression, southern Norway.*

Dwarf mongooses keeping lookout from a termite hill, Samburu Game Reserve, Kenya.

Above: (left) *A male olive baboon asserts his dominance by displaying his teeth, Kenya;* (right) *vampire bats roosting in a cave, Costa Rica.*

Below: *Naked mole rat queen suckling her young in their nest chamber, Kenya.*

Above: *Leaf-cutter ants transporting sections of cocoa leaf which carry riders. These will ward off any parasitic flies that might otherwise lay their eggs on the leaf and so gain entrance to the nest, Trinidad.*

Below: *The formidable jaws of a soldier leaf-cutter ant, Costa Rica.*

Above: *A springbok displays by 'stotting', jumping with stiffly held legs, Kgalagadi Park, South Africa.*

Below: *Male pronghorn antelope signal a silent and far-reaching warning by expanding the white rosette of fur on their buttocks, South Dakota.*

Above: *Semipalmated plover using a broken-wing display, feigning injury to lure intruders away from their egg-filled nest, Icy Bay, Alaska.*

Below: *Three-wattled bellbird male calling and displaying from perch in cloud forest, Costa Rica.*

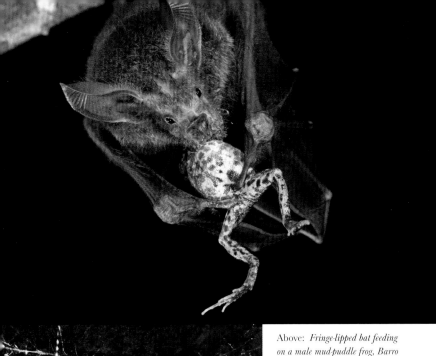

Above: *Fringe-lipped bat feeding on a male mud-puddle frog, Barro Colorado Island, Panama.*

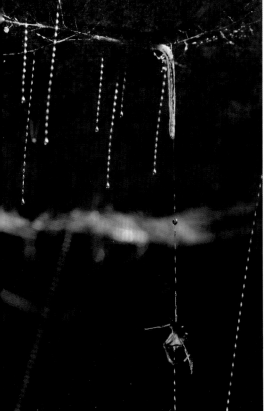

Left: *A fungus gnat larva with its prey caught by a silk thread loaded with beads of glue. Glow-worm Cave, North Island, New Zealand.*

A glow-worm on a stem of grass, displaying its bioluminescent abdomen, England.

Above: *A wild almond tree in which, at night, thousands of Luciola fireflies synchronously flash their green lights, Luzon, Philippines.*

Below: *A firefly squid emitting light from photophores that serve as counter-illumination camouflage, Toyama Bay, Japan.*

Above: *A male bee-eater offers a wasp as a courtship gift, Hungary.*

Below: *A sandwich tern presents his mate with a fish during courtship, Caithness, Scotland.*

Left: A female dance fly feeds on the gift of a march fly brought by a male, as they mate, Wiltshire, England.

Below: A male nightingale in full song, Lincolnshire, England.

Above: *Superb lyrebird male in courtship display, Victoria, Australia.*

Below: *A male Lawes' parotia bird of paradise dances on his display ground while a female watches from a branch above, judging his performance, Papua New Guinea.*

Above: (left) *A Temminck's tragopan male displaying with unfurled lappet;* (right) *a male Victoria riflebird displaying with raised wings, Queensland, Australia.*

Below: *A male blue bird of paradise performing his upside-down display, Papua New Guinea.*

Above: (left) *Male topi standing alert on a mound on the frontier of his court, Masai Mara, Kenya;* (right) *Gunnison sage grouse male displaying at a lek, Colorado.*

Below: *Satin bowerbird male arranging blue plastic ornaments and feathers at a bower, watched over by two females, Australia.*

A male MacGregor's bowerbird tidies up the circular runway around his maypole on which he will dance, Papua New Guinea.

Above: *Nudibranchs mating, Lembeh Strait, Indonesia.*

Below: *A blackfoot tarantula male secures the female's fangs and bends her backwards as he pushes his sperm-filed palps into her reproductive slit, northern Arizona.*

Above: *A male four-spot orb-weaver spider approaching a much larger female in order to mate, Peak District, Derbyshire, England.*

Left: *The tiny male and gigantic female golden orb weaver spider on their web, Colombia.*

Above: *Male southern elephant seals fighting for breeding rights, St Andrew's Bay, South Georgia Island.*

Below: *An African bull elephant mounting a female, Masai Mara, Kenya.*

Above: *A lioness rejecting a male's advance, Masai Mara, Kenya.*

Left: *An adult northern royal albatross with its chick, Otago Peninsula, New Zealand.*

Right: *Black darter dragonfly pair in a mating wheel on heather, Surrey, England.*

Below: *Heliconius butterfly males attempting to mate with a female emerging from her pupa, Costa Rica.*

mounds, and on sandy soils they are almost completely underground. In one small area in Nigeria, each nest is a cluster of towers and minarets grouped around a central spire that may rise six metres high. This particular version contains within it a cooling device of unsurpassed elegance.

The main part of the nest lies below ground level beneath the towers. Two metres down, there is a huge circular cellar, three or four metres across and a bit less than a metre high – quite big enough for a human being to crawl into. Its hummocked floor is studded with shafts which descend a further four metres or more to reach the water-table. Few termites are found down here. Those that do come are small pallid workers, inching their way in long columns across the floor and down the shafts to collect the moist mud that is needed for further building work. The scale of the structures around them is so huge they look like teams of porters marching over a range of hills to a mine.

In the centre of the floor stands a massive clay pillar. This supports a thick earthen plate which forms the ceiling of the cellar and carries above it the central core of the nest with its tiers of nurseries, fungus gardens, food stores and, of course, the royal chambers where the king and queen live. It is on the underside of this plate that the bellicose termite constructs its most spectacular architectural invention. Rings of thin vertical vanes up to fifteen centimetres deep, centred around the pillar, cover the ceiling like a diffusing grid in a giant lamp-shade. In fact, they are not separate rings but one continuous spiral with its coils only a couple of centimetres apart. Its lower edge is fretted with holes, like lace. Its sides are white with encrustations of salt. This delicate structure made of hard dried mud, absorbs moisture through the ceiling from the nest above. This evaporates from the surface of the spiral. As it does so, the white salts are deposited. But, much more importantly, the process of evaporation cools the air around and makes the cellar the coldest place in the entire building.

Heat generated by the fungus gardens and termites in the main part of the nest above the basal plate causes the air to rise through the passageways and chambers until it reaches the large spaces in the upper part of the nest and within the towers. From this loft, a number of flues run downwards, close to the outside walls, leading eventually past the edge of the basal plate and into the cellar. As heat continues to be produced in the core, so air in the loft flows down these flues, drawn by the coolness of the cellar beneath. The external walls of the flues are constructed from a particularly porous earthen material, in places pierced transversely with tiny galleries that end very close to the outer surface so gases can easily diffuse through them. As the stale air travels slowly through the flues, so carbon dioxide flows out and oxygen flows in. By the time the air reaches the cellar, it has been refreshed and cooled. With this ingenious structure, simple in principle but most complex in its architectural intricacies, the bellicose termite keeps its fungus beds permanently between 30°C and 31°C, exactly the temperature its precious fungus requires.

If the dimensions of such a nest were translated into human terms, with each worker termite being considered the size of a human being, then this amazing fortress would stand two kilometres high. If we were to start on a building of such magnitude – which we have not yet attempted – it is easy to imagine the army of architects and engineers, the volumes of plans, the batteries of computers, the regiments of construction machinery we would require. Yet these million or so termites build their equivalent working in a coordinated way in total darkness, each blind, tiny-brained insect knowing exactly where it has to place its pellets of mud to produce nurseries, supporting pillars, living chambers, gardens, flues, defensive walls – and that extraordinary spiral cooling vane.

As with so many of the buildings constructed by animal architects, we really have very little idea how they do it.

SEVEN

Living Together

Acommodious and secure home can only too easily attract the attentions of others. Whether the rightful owners like it or not, lodgers may move in. The spacious empty flues of the termite fortress are just the kind of accommodation that suits dwarf mongooses. A medium-sized hill has more than enough space for a family of a dozen or so and if rooms are vacant, dwarf mongooses seldom live anywhere else.

Several birds, including parrots and woodpeckers, also favour termite mounds. The golden-shouldered parrot of Australia nests only there. It frequently tries to do so in one of the great blades constructed by magnetic termites and starts nibbling a hole in the flank. But then, before the tunnel is long enough to be widened into an egg chamber, the bird – to its obvious bewilderment – breaks through on the other side and has to abandon the project. It has much more success with its usual choice of a cone-shaped mound. There it excavates a commodious nest chamber in the crumbly earth of broken galleries, which it leaves unlined with any other material.

Some species of termites wage a persistent war with such intruders, doggedly trying to repair the entrance hole at night, even though the birds, with equal persistence, demolish the reconstruction in the morning. Usually the birds get their way, but on occasion the termites succeed and a nestful of chicks are entombed. Other termites, however, seem to become reconciled to losing part of their property and wall off the broken galleries, so that the nest chamber becomes a separate and self-contained apartment, completely partitioned from the main building.

The tunnels dug by American gopher tortoises are also commandeered by squatters. Snakes slither in to cool off in their shade when the sun gets too hot. Burrowing owls, rather than burrowing for themselves, move in too, sitting in a proprietorial way beside the entrance and glaring in seeming outrage when the rightful owner lumbers in.

In New Zealand, shearwaters returning after months at sea to nest in tunnels on the tops of cliffs often find that in their absence their holes have been taken over by primitive lizards, tuataras. Once established, a tuatara becomes the permanent year-round caretaker, keeping the hole clear of blockages so that when the birds return the next season, all they have to do is to clear out the nest chamber at the far end. A tuatara will eat shearwater eggs and chicks but, considerately, it never takes those belonging to its landlord that lie in the far end of the shared tunnel.

A hermit crab makes its home in the vacated shells of whelks, winkles and other molluscs. It slips its soft, curling rear-end into the spiral tunnel of the shell and withdraws inside, closing off the entrance when necessary with its armoured claws, like a boxer shielding his face with his fore-arms. But the fit between body and shell is not so tight that there is no room for anyone else. One large species of hermit which lives in whelk shells is regularly forced to accept a rag-worm as a lodger. The worm, once it has gained entrance, seldom if ever leaves. When the crab, trundling about on the sea floor dragging its home with it, finds something to eat and starts to shred it with its claws, the rag-worm sticks its head out and daringly snatches fragments directly from the crab's scissoring mouthparts.

But sometimes the lodger is useful. Another species of hermit crab habitually carries a large anemone on top of its shell. All anemones have weapons in their tentacles, microscopic capsules that, when stimulated, release poison darts. Small fish and even octopus take care to avoid them, so although these predators eat crabs when they get a chance, they keep clear of one that has an anemone as a companion. The hermit clearly values the anemone's protective presence, for it makes sure it retains it. As the crab grows, it has to transfer into more commodious premises. Having made the move, it holds up its vacated shell with its pincers and, with its other legs, detaches the anemone and replants it on top of its new home.

A crab in the Pacific Ocean uses anemones as deterrents in an even more direct way. It carries one in each claw, like a pom-pom dancer (hence its name, the pom-pom crab). If a fish interferes with the crab, it gets a fistful of stinging tentacles in the face. This crab has become so reliant on its armament of anemones that its pincers have lost much of their strength. They are no longer sufficiently strong to tear apart its food. To do that, it has to use its next pair of legs. If a crab loses its anemones, or they die, it will try and steal replacements from another crab; if it succeeds, both crabs will then gently rip apart their anemone, holding each half in its front claws. The obliging anemone will then regenerate into two separate but identical organisms, restoring the status quo.

Some species of anemones themselves play host. Small clown fish spend most of their lives lounging among their writhing sting-studded arms without being injured in any way. It seems that the stimulus that causes the stings to discharge is, at least in part, a chemical that is in the mucus covering of most fish. The clown fish's skin lacks this substance, so it is not attacked. Although young clowns when approaching an anemone for the first time do so gingerly, brushing only one or two tentacles very gently, their confidence builds up quite quickly until they are able to plunge around the anemone with impunity. Even when the anemone completely withdraws its tentacles and closes up, the clown fish may remain inside it without coming to harm.

The fish without doubt gets considerable protection from the anemone. It seldom strays far from it and whenever danger threatens it rushes back and dives into the anemone's protective arms. The benefit to the anemone from the arrangement is less clear. The clown occasionally cleans off dead tissue and waste from its host. Its very presence may also encourage other small fish to swim close to the anemone which then has a chance to catch them and so get a sizeable meal. One species of clown, however, certainly helps its host in an active way. Some butterfly fish on the reefs of northern Australia and New Guinea are, like the clowns, unaffected by the anemone's stings and although they feed mostly on coral, they will bite lumps off the anemone if they get the chance. But if they approach one that has a clown fish in residence, the little clown comes out chattering its bared teeth and threatening it with such vigour and confidence that the intruder is driven away.

No bargain, of course, is knowingly struck between such partners. Each is exploiting a situation to its own maximum advantage and it is often difficult for a human observer to decide which is getting the better of the arrangement. The Australian green tree ants that glue their leaf nests together with silk from their grubs, eat caterpillars. However, the caterpillar of the common oak blue butterfly has on its back a little nipple which, when stimulated, produces drops of sweet sugary liquid called, somewhat flatteringly, honey-dew. Its skin is also liberally dotted with other tiny glands that secrete amino acids. Both these substances are eagerly eaten by the green ants and instead of tearing apart this kind of caterpillar, they look after it with great solicitude. They build a small shelter for it in which it spends the night. They accompany it when it marches out in the morning to feed, running excitedly around it and even clambering on its back. If a predatory wasp or a spider comes near, they drive it away with squirts of formic acid. And they regularly milk the caterpillar, stimulating its honey-dew gland and combing amino acid particles from its skin. It seems clear that the ants are exploiting the caterpillar just as a farmer exploits a cow.

But the caterpillar also benefits from the arrangement. If common oak blue caterpillars are put in trees where there are no ants, they are quickly picked off by predators of one kind or another. In experiments to assess these risks, several hundred such caterpillars were used and not a single one survived. So is the caterpillar the passive beneficiary of the ants' enterprise or does it take a more active part in creating the relationship? One feature of its anatomy gives a clue. It has a pair of little plumes on its back which give off a scent when erected. Furthermore, it makes a low rumbling noise which you can feel as a vibration if you take the caterpillar on your finger. Both these signals attract ants. They may also act as identifying signals that tell the ants that the caterpillar is the kind that should not be eaten but cherished, probably because it smells like one of their brood. So maybe the caterpillar should be regarded not so much as a cow kept and exploited by farmers, but as a fat emperor who has recruited a guard of heavily armed warriors whom it pays with daily rations of food.

An evenly balanced relationship such as this, however, can easily tip into one where one partner exploits the other and gives nothing whatsoever in return. The same green tree ants also associate with the caterpillar of the moth-butterfly, a close relative of the common oak blue. It does not have a soft skin like the oak blue. Instead, it is covered from head to tail by a brown oval shield. It is neither cow nor emperor. It is a tank. This strange creature makes its way into the leaf-nest of the tree ants, moving slowly with the edges of its carapace held so close to the surface of the leaf over which it is crawling that the ants cannot get beneath it to attack its soft parts. Nor can their jaws make any impression on its smooth horny surface. Eventually the caterpillar reaches the part of the nest where the ant grubs lie. As it moves alongside one, it suddenly tilts up the side of its shield and clamps it down again with a luckless larva trapped within. Then, in the safety of its impregnable shelter, it slowly eats its catch.

The intruder spends the rest of its caterpillar existence within the ants' nest feeding in this way. The ants can do nothing to stop it, nor can they eject it – again, it is most likely that the caterpillar has evolved a set of chemical signals that are nearly identical to those of the ants. They literally do not sense anything wrong.

Here the caterpillar pupates and here, surrounded by ants, the adult butterfly emerges. Now might seem to be the time that the aggressive and well-armed ants could get their revenge. But even though the butterfly no longer has an armoured shield, it still has the same chemical camouflage, and for added security, it has a different form of protection. It is called moth-butterfly because its wings, body and legs are all covered with loose white scales. These slip off when the ants attack it and so clog their jaws and antennae that the butterfly is able to elude them and make its escape into the outside world.

If there is a great disparity in size between the two participants in such a relationship, each is bound to have a very different attitude to its partner. For the smaller, the vast body of its host is simply another environment with its own rewards and hazards like any other. To the bigger, its tiny guests may be so small that they are scarcely noticeable. Sometimes they may be helpful and even worth encouraging. But once the relationship has been established, consciously or unconsciously, willingly or unwillingly, it may become increasingly intimate and have profound effects upon both.

The three-toed sloth, hanging from a branch in the South American rainforest, for most of the time apparently fast asleep, seems so careless of its own personal toilet and so disinclined to any kind of violent action that there is little to prevent any animal that wants to do so from living within its coat of coarse shaggy fur. And many organisms do. Algae grow on the filaments of its long outer hairs. These have a form that is quite different from other hair with surface scales beneath which the microscopic algal cells lodge. Whether these structures developed specifically as accommodation for the algae, and if they did, what benefit the sloth might get from them, no one knows. There are also large numbers of small moths scuttling about in the fur. A single sloth may have as many as a hundred of them. It was once believed that the moths and their caterpillars fed here, grazing on the algae. Now it has been shown that they do no such thing. The female moths lay their eggs on the sloth's dung which it deposits in special middens on the ground. There the caterpillars feed and pupate. The adult moths, when they emerge, use the sloth's coat as a moving carpet to transport them from one breeding ground to the next. It probably also provides them with the opportunity during the journey to mate.

A little mouse that lives in the forests of Costa Rica regularly carries as many as a dozen beetles in its fur in a similar way. They cling to its ears and neck and crawl all over its face. They are seldom found anywhere else except with the mice. It used to be thought that these also fed on their host, perhaps sucking its blood, for they have very large jaws. Mysteriously, however, those mice with the largest number of beetle passengers, far from being debilitated or anaemic as might be expected, seemed particularly healthy. In fact the beetles' jaws do no more than give their owners a firm grip on their host as it runs through the forest at night. They feed during the day. At that time the mouse is in its burrow and the beetles leave its fur to hunt for the fleas that abound in the mouse's nest. Because the beetles keep down the number of fleas, the more beetles a mouse has, the healthier it is likely to be.

But some animals that frequent the bodies of bigger ones do provide a genuine valet service. Many of the big animals of Africa – eland and buffalo, warthog and rhino – are attended by oxpeckers, birds that belong to the starling family. A giraffe may have a flock of several dozen of them as its regular attendants. They scuttle over its body, picking off fleas and ticks and maggots, clambering into its ears, pecking around its eyes, probing beneath its tail. They are so at home on the giraffe's body that they perform their courtship displays there. This relationship started so far back in evolutionary history that the oxpecker has now acquired several adaptations which help it in this particular way of life. Its beak has become flattened so that it can put its head on one side, probe deep between the hairs of the coat that lie flat and close to the skin, and give a tick the firm strong tug needed to pull it loose. Its claws are particularly long so that it can hang on even when its host is giving it a rough ride. And its tail is more like that of a woodpecker than a starling, being stiff and short so that it can be used as a prop when the oxpecker clambers up the flank of an eland or the towering neck of a giraffe.

The service provided by these birds is a very real one. Without them a warthog could not possibly remove a tick from its ear or a buffalo rid itself of a maggot at the base of its tail. Because of this, their hosts allow them great freedom to wander where they will, into all the clefts and crannies of their bodies. But the birds are not an unqualified benefit. Blood is a major part of their diet, for the stomachs of the ticks they swallow are full of it. And they are not always content to take it second-hand. If their host has a wound or a sore, the oxpeckers will cling beside it, pecking at it and supping the blood when it begins to run. By doing that, they are not improving their host's health, they are damaging it, keeping a wound open long after it would otherwise have healed.

In the Galapagos, finches attend to giant tortoises. They alight in front of one of them and hop up and down in an exaggerated fashion. If the tortoise feels the need to be cleaned, it signals its acceptance by craning its neck upwards and stiffening its legs so that its huge shell is lifted clear from the ground. In this position, all the more intimate parts of its skin where something unpleasant and irritating may have lodged are as fully exposed as they can be. Immediately, the finch flies on to the tortoise, inspecting its neck and climbing up its thighs while the tortoise stands quite motionless with that air of frozen patience adopted by a human being having their hair cut.

The same sort of services are also available in the sea. Huge sunfish come to the surface of the water and float on their sides so that gulls can come down and remove fish-lice from their flanks. Phalaropes have been seen doing the same thing for whales. On coral reefs there are special places that are recognised by fish as cleaning stations. Here a resident staff of small wrasse and shrimps is always on hand. When a big grouper or a parrot fish cruises in, a cleaner wrasse, a slim little fish wearing a vivid uniform of blue and white stripes, dances in front of the new arrival with a bobbing motion. The grouper now hangs in the water, holding open its gill covers and mouth, often with its body tipped more vertically than horizontally, sometimes head-up, sometimes head-down, in a posture that signals its willingness to have its toilet attended to. The little wrasse swims in and fusses all over its client, trimming off pieces of dead skin, snipping away infestations of fungus, boldly venturing right into the huge jaws and coming out through the gaping gill covers. Many fish return to these stations every few days for servicing and even though the cleaners can deal with as many as three hundred customers in six hours, there may still be queues awaiting their turn.

The organisms removed by these cleaners are mostly fish-lice, specialised crustaceans that spend all their adult lives on the bodies of fish. Some feed by scraping tissue from the surface of the skin. Some suck blood and some burrow so deep into the fish's body that only their tails show to the outside world. These creatures are not innocuous passengers like the mouse's beetles or the sloth's moths. They do not provide a service of any kind whatsoever. They feed on the flesh of their hosts and give nothing in return. They are parasites.

Many different kinds of animals have taken up the parasitic way of life. Some arthropods took to it early in the terrestrial history of their family. They became ticks. Their eight legs are short and stout and their jaws are particularly strong. They have special sense organs on the ends of their front legs that can detect changes in humidity and odour and so help them to find a host. A hungry tick looking for a meal climbs up a stem or a leaf in the undergrowth and waves its front feet in the air, ready for the moment when some warm-blooded animal brushes by and it can climb aboard. That may seem a fairly remote possibility in many circumstances. But adult ticks are patient animals. They can wait as long as seven years between meals.

When the happy moment arrives, the tick scrambles on to its host, makes its way through the hair to the skin and there cuts a little incision with its pincers. Holding on with its teeth, it inserts its grooved snout, covered with backward-pointing hooks, up which it sucks its host's blood. It remains there gorging itself for several hours during which time it may swell from the size of a rice grain to a fair-sized button. If it is mature, it will then drop off to breed.

Mites are related to ticks. Nearly all of them are very small indeed and many species have taken advantage of their minuscule dimensions to take up residence in the tiniest pockets and apertures on the bodies of other animals. One species lives in the small pits on the wing-cases of Trinidadian harlequin beetles and nowhere else. Another is found only on the feathers of a nightjar's wing – and only in the white ones, never in the brown. Whole colonies reside in the ears of moths. They use separate parts of the ear for egg-laying, for stacking their refuse and for feeding, which they do by sinking their mouthparts through the moth's cuticle and sucking its body fluids. But, obligingly, they occupy only one of the moth's ears. If both were blocked, the moth would be deaf and easily fall victim to such predators as bats – and that would be as disastrous for the mites as it would be for the moths. Uniquely among the arthropods, two whole families of mites have even made the return journey from the land into the sea. Over one thousand species of mite now live at various points in the ocean, from the tidal zone to the abyssal depths, 7,000 metres down.

Insects have taken to parasitism in a particularly extensive and diversified way. Indeed, one entomologist has calculated that a tenth of all species of animals in the world are parasitic insects. Even insect families whose members nearly all live independent lives may contain one or two which have turned to parasitism.

Several species of fly live permanently on the skins of birds and mammals, crawling about between the hairs or feathers and sucking their host's blood. Claws are very valuable to them as such flies need to keep a firm foothold, and they are very much enlarged. Wings, however, would be a hindrance in such a situation and theirs are greatly reduced in size. So these have become flies that are flightless.

Another insect family, the fleas, are all parasites and all of them have lost their wings entirely. They still need to travel from one host to another. Lacking wings they cannot fly, but they can leap and they do so phenomenally. They use a device based on a structure in their flanks that once served as a hinge for their ancestors' wings. It is made of an elastic substance called resilin which, as the flea gets ready to jump, is slowly compressed and then locked into position. When it is released, there is an audible click and the hind legs straighten with such force that the flea shoots into the air with a leap that, in comparison to its size, is the equivalent of a human jumping over an office block. A cat flea can easily jump a full thirty centimetres when searching for a new host. Some fleas live most of the time in the nests and burrows of their hosts and only hop on to their bodies when they need to feed or move to another site. Others, however, have become permanent passengers.

Yet another insect family, the lice, are also all wingless. They probably originated, like the fleas, as nest-living blood-suckers, though now all of them take up residence on their hosts' bodies.

Even though the claws and jaws used by all these creatures to attach themselves to their hosts are large and strong, they are seldom totally effective. As long as a parasite lives on the outside of its host's body, there is always a chance that it will be dislodged. Big animals can scratch and rub themselves. Sea snakes, lacking legs, tie their long bodies into slip-knots which they work down their length from head to tail so that one fold, as it passes over another, rubs off parasites that may have fastened themselves to the snake's skin. Human beings and monkeys can pick off a tick with their fingers. And if the parasite has lodged itself somewhere that the host itself cannot reach, then a cleaner may root it out. But many parasites find a refuge where nothing can reach them – inside the bodies of their hosts.

Pearl fish are pre-adapted to doing so. They are long, slim and scaleless. Most live in clefts in rocks on the sea floor which are so narrow that they are unable to turn round in them, so they enter tail-first. But one species which feeds in open water at night spends the day sheltering inside the mantle cavity of a pearl oyster. It seems to lodge there with impunity, but occasionally, for some reason, one dies there. Then the oyster lays down mother-of-pearl all over the body cementing it to one side of the shell, which is why the fish was given its name. Another species, known as the messmate, chooses not an oyster but a sea cucumber for its home. This is a sausage-shaped relative of the starfish that lies on the sandy floor of the sea inoffensively sucking up sediment. The messmate, when it is young and small, is able to insert its head into the sea-cucumber's anus and wriggle its way into the spacious water-filled body cavity. When it grows older and bigger however, it cannot manage this. Instead, it has to insert its sharp pointed tail into the anus and then, holding its body in a spiral, revolve so that it winds itself in, like a cork-screw. Although some species of messmate use the cucumber as nothing more than a shelter, others munch their host's internal organs which, fortunately for both of them, the cucumber is able to regenerate continuously.

The animals which, by their very shape, are best suited to the interior life are those long, spineless, legless creatures we call collectively worms. We are most familiar with those whose bodies are divided into ring-like segments, such as earthworms. This group also includes the leeches which drink blood when they get the chance and have a powerful sucker at either end of their bodies to enable them to hang on to an animal's skin while they are doing so. They do not venture inside their hosts. But there are also roundworms that have no segments and look like writhing pieces of thread or string, and flatworms that are shaped like ribbons or leaves and glide over the ground or undulate through water. These two groups have taken to internal parasitism in a wholesale way.

Although we hardly ever see roundworms, they are among the most widespread and diverse of all animals in the world – there can be four and a half million individuals in a square metre of mud, billions in a hectare of soil. Taking free-living and parasitic forms together there are probably half a million or more different species, a membership exceeded only by the insects. Every vertebrate that has been thoroughly examined, whether fish or amphibian, reptile, bird or mammal, has proved to be the host of a roundworm of some kind. They burrow within the body of the animals they infest, boring through the tissues to reach the site they most favour and there they live by absorbing blood and other body fluids. They form cysts in muscles and bury themselves in glands. They dwell within eyeballs and they so distend the human body that legs swell to the size of those of elephants.

Parasitic relations of the flat worms, the flukes, live largely on blood. Some suck it from the exterior of their hosts, squirming their way into nostrils, vents and mouths. But many others go far deeper, living inside the liver and other parts of the body, maintaining their position with circlets of hooks around their mouths. Tapeworms, shaped like long ribbons that can grow to several metres in length, specialise in feeding on the soup of digested food that swills down the gut of their hosts. They lack virtually any sense organs of any kind. They have little need of them. They do not even have a gut of their own. They allow their host to do all the digestion necessary. All they do is absorb the resultant fluids directly through their skins.

Some of these animals cause their hosts little more inconvenience than robbing them of a proportion of their food and their flesh. But many severely damage their hosts' health, either by physically injuring their tissues and organs as they make their way around within, or by excreting waste products that poison their host and cause fevers and ultimately death. But they themselves have attained the most secure of lives, hidden away within another animal's body where no enemy can reach them and surrounded by a supply of food that never ends as long as their host survives.

Their main problem in life is ensuring that their young manage to attain an equally ideal position within another body of the same kind. That will not happen if the eggs remain alongside the adults, for few host species are cannibals, eating the bodies of their own kind. The eggs must somehow leave the host, and the easiest and most obvious way to do that is with the host's droppings. But even this cannot be the solution for all, for only a minority of mammals and birds are so insanitary that they become soiled by the dung of their fellows.

The solution to the problem provided by evolution is to recruit a different kind of host. A tapeworm living in the gut of a cat uses a mouse. Its eggs are shed with the cat's droppings. These may fall among grain or some other potential mouse food where they may be inadvertently consumed by a mouse. The eggs, inside the mouse's body, hatch into larvae which migrate into its liver. There they form cysts and multiply into large numbers of yet a different form of larva. If the mouse is then caught and eaten by another cat, then another generation of tapeworm has succeeded in finding a new home.

Such a circuit is, of course, a very chancy one. Since the mouse is not seeking cat droppings, the vast majority of them will remain untouched. Even if the mouse does happen to eat a smear of one, the parasite's future is still far from assured. If the mouse is caught, not by another cat but by a dog or a fox or an owl, then the larvae within it will live no longer. Maybe only one in a hundred thousand will have the good luck to reach another cat. For this reason, internal parasites produce their eggs in astronomical numbers. A mature tapeworm, living in a human gut and needing to get its larvae into the flesh of a pig in order that another pork-eating human being will be infected, may shed a million eggs a day and over its lifetime produce as many as seven thousand million of them.

A few of these internal parasites have developed ways to improve their chances of completing the hazardous connections in their life-cycle. In northern Europe, many small birds, such as flycatchers and thrushes, carry flukes within their gut. The parasites' eggs fall to the ground in the birds' droppings where they may be eaten by a grazing snail. Inside the snail's body, they hatch into small actively-swimming larvae that bore their way through the gut wall and into the liver. There they reproduce themselves and form little mobile cysts which make regular journeys every morning into the snail's tentacles. These are normally thin, but when the parasite forces its way into them they become thick and club-like. Not only that, but the stretched wall of the tentacle becomes so thin that it is transparent and the parasite within is easily seen. It is brilliantly coloured, banded with yellow, orange and dark brown. To make itself even more conspicuous, it pulsates. The presence of the parasite for some reason also changes the snail's behaviour. Instead of returning to the safety of the leaf-litter soon after dawn, as uninfected snails do, it remains out in the open for much longer. The throbbing coloured bands within the swollen tentacles quickly attract the attention of foraging birds. Perhaps it is their resemblance to caterpillars on which many birds regularly feed. Whatever the reason, the birds flutter down, peck the tentacles off the unfortunate snail and swallow them. Once again, a new generation of parasites has managed to reach the same kind of safe home as that in which its parents flourished.

Nor is such a life, spent alternating between two hosts, the most complex followed by internal parasites. Some have not two hosts but three. Flukes in the livers of rabbits shed their eggs, as so many parasites do, in their hosts' droppings. These are inadvertently eaten by land snails. There they form slime balls which the snail extrudes from the small hole on the side of its head through which it breathes. These balls are eagerly eaten by a particular species of ant. Each contains several thousand larval parasites and once inside the ant's stomach they spread throughout its body. Some make their way to the ant's brain and so interfere with its workings that the ant starts to behave in a very eccentric way. Instead of returning to its nest at night, it locks its mandibles on to the tip of a leaf. There it remains until morning. Only when the sun is well up and the temperature begins to rise will it release its hold and return to its nest. But it seldom survives long enough to do so. Most such ants, hanging in exposed positions, are inadvertently eaten by other rabbits nibbling the grass in the early morning. Once inside a rabbit's body, the larvae develop into flukes and the bizarre cycle, once again, has been completed.

So it is that few big animals are truly solitary independent individuals. A buffalo standing in a swamp stolidly chewing the cud is not alone. Oxpeckers cling to its flanks. Ticks are boring into its hide. Leeches may have fastened on to it when it went to drink and now lie within its mouth attached to its lips. Tapeworms, hidden from view, may be trailing through its convoluted gut, roundworms encysted in its muscles and flukes moored in the veins of its liver absorbing its blood. All these creatures are robbing it of sustenance. But there are still others, even smaller, which are providing it with food and without these it would starve. Microscopic organisms are swarming in the compartments of its stomach, helping it to break down the cellulose in the plants that it has eaten which otherwise it could not digest.

Most large animals, in fact, are not the single individuals they seem to be. They are walking menageries, whole communities of different species which, in their various ways, are committed by evolution, for better or for worse, in sickness and in health, to live together.

EIGHT

Fighting

The necessities of life, in this harsh world, are often in short supply. An animal needs food, territory and, when the time comes, a mate. If it is to have these things, it may be forced, on occasion, to fight its own kind to get them.

A grizzly bear is the most powerful animal in North America. It can kill a moose or a human being. It fears nothing. Since it leads a largely solitary existence, it seldom needs to make allowances for others. Nor does it. If it finds the small carcass of a deer lying on the chill Alaskan tundra, it will feast with fearsome strength, ripping open the belly, tearing the flesh from the bones. But the smell of carrion travels far through the clear cold air. It may well reach the nostrils of another bear a couple of kilometres away. It too will want a share. If it is much smaller, it is likely to loiter at a respectful distance, waiting for the first to leave and hoping that there will be something left. But if the newcomer is of comparable size or even bigger, it may not feel inclined to defer. And then there may be real trouble. Snarls and growls lead to bites and ferocious swipes from huge paws as each animal fights for a share. If the carcass is small and food is scarce, then the quarrel may be a critical one and fought with little quarter given. In the end, straightforward fear and injury will dictate defeat and sheer strength will win the day.

Even animals that live habitually in groups may sometimes quarrel violently. When a big animal is killed on the open plains of Africa, vultures of several different kinds fly in to clear up the carcass, pulling out the guts, tearing off the skin, ripping away what fragments of muscle the killers themselves may have left. With so much food concentrated in such a small area, there are bound to be many claimants and therefore many squabbles as each individual bird argues noisily and vigorously with others in order to get a share.

It is hardly surprising that bears and vultures should squabble in such situations. But even the most simple and apparently inoffensive animals may be driven to violence. Sea anemones might appear to be placid creatures, stuck to a rock, groping through the water with gently waving arms. But they too may have to compete for food. Unless they occupy a site where the water moves at a reasonable speed and brings with it a good supply of nutriment, they cannot flourish. So they too have duels.

A well-fed beadlet anemone, clinging to a rock on a European shore, buds off daughters so that often several genetically identical individuals sit alongside one another amicably waving their tentacles together. But the richness of the food supply may also attract another individual of the same species. The beadlet exists in two colour-forms, brownish-red and green, and it is often the red that is more active and likely to move. It travels by making slow undulations of its base, advancing its margin millimetre by millimetre and taking an hour to travel one or two centimetres. But it moves purposefully and eventually it arrives. To our eyes, the two organisms look exactly the same apart from their colour, but the resident anemone is immediately aware that the tentacles that have touched it are not those of one of its sisters but are genetically different. They belong to a stranger. If it is to maintain its position, the resident must fight.

Its weapons are the microscopic harpoons laden with poison that are each coiled and enclosed within its own cell. Some are placed along the length of the tentacles. Others are grouped in dense batteries in a ring around the crown of tentacles. Normally they are used to capture and paralyse small fish. Now they must be deployed against a rival. The resident quickly withdrew those tentacles that were touched by the intruder and now it inflates part of the outer ring so that batteries of harpoons point towards the stranger. As the alien arms continue to make contact, so the tiny threads are discharged. The barbs in their tips inject poison and stick fast to the skin of the intruder, pulling it off in strips. The intruder too has harpoons and returns fire. Eventually, one or the other surrenders, retracts its entire crown of tentacles, closes up into a rounded blob and creeps away.

Slugs also become extremely savage. They may be unaware of one another's existence as they glide towards a suitable site to lay their eggs until the sensory tentacle of one, peering forward, touches the other. Immediately that happens, they recoil and prepare for battle. Each inflates the hood-like mantle that partially covers its head and rears up to expose the gaping mouth on the underside of its body. This has a long rasp-like lower jaw on to which the upper jaw repeatedly closes like a guillotine. One now lunges forward and delivers a slashing bite on the flank of its opponent and the two begin to exchange blows in what may be called, with double accuracy, a slugging match. Eventually one, with its sides badly torn, begins to retreat. Its wounds exude a poisonous mucus and if its opponent gets a mouthful, it finds it so unpleasant that it breaks off the engagement to wipe its mouth on the ground. That gives the vanquished a chance to escape. Some species discharge a great puddle of slime which temporarily diverts the victor's attacks. But often neither of these things distracts the winner, who gives chase, striking repeatedly at the loser's body with such effect that it may eventually die.

Straightforward fights of this kind, in which contestants are likely to be injured or killed, are clearly not the best way for animals to resolve their differences. Both parties will benefit if, somehow, quarrels can be decided by debate rather than by blows, if contestants are able to assess one another's strengths and accept the probable outcome of a fight without putting the issue to the test.

Territorial disputes usually go through several stages before they come to blows. They start with a declaration of ownership. Birds make this with sound. A pair of tawny owls, once they have set up home together, dissuade other owls from moving in by hooting from different points around their territory throughout the year. A skylark, staking its claim to a nesting site in the grass tussocks of a meadow and lacking any tree in which to perch, trills its proclamation of possession on the wing high in the sky.

Mammals also use their voices to claim territory. Indris, the largest of all living lemurs, sit on branches in the rainforest of

Madagascar, lifting their heads to trumpet long series of calls at one another to establish their rights as landlords. Teams of howler monkeys in South America and families of gibbons in South-east Asia sing in chorus, morning and evening, and often provoke an equally defiant claim from their neighbours. The long whooping yodels of a pair of gibbons, echoing over the tops of the trees, is one of the loveliest and most evocative sounds that any traveller will hear in the forests of Borneo. To other gibbons, however, it is a clear warning that trespassing will not be tolerated.

The shrilling sounds of cicadas serve the same purpose. These insects have special circular membranes on either side of their abdomen which make a pinging noise as they snap back and forth. When they are oscillated at very high speed they produce one of the loudest of insect calls. Grasshoppers sing rather more modestly by rubbing their notched thighs against the edge of their wings. Even a few butterflies duel with sound. The male hamadryas butterfly of South America claims a sunlit patch on a boulder or a tree trunk in which he can display to attract a female. He sits there posturing with his wings, but if another male arrives he flutters into the air and makes a loud clicking noise using microscopic structures on his wings. The two may circle one another in an aerial dog-fight, sounding off their tiny machine guns until one or the other gives up. A butterfly's vision is not very acute and a possessive hamadryas will challenge other butterflies of quite different species if they come near his patch of sunlight, and sometimes, valiantly if misguidedly, launch an attack on a small bird.

Smell, in some ways, is an even better medium in which to express a notice of ownership. For one thing it lasts longer. Badgers use their dung for this purpose and defecate regularly in latrines strategically placed around their territorial frontiers.

Dogs use their urine, and several species of deer produce a smelly paste from glands beneath their tail or on their cheeks, which they smear on tree trunks or leaves.

The ring-tailed lemurs of Madagascar produce scent not only from glands beneath the tail but from others on the wrists. As they patrol their territory, they mark particular trees. But on occasion these smelly notices need reinforcement. If one group encounters another near their common frontier, the males will start a stink fight. They rear up on their hind legs, bring their long tails forward between their thighs and draw them over the glands on their wrists so that the long fur is impregnated with scent. Having done that, they drop on all fours and with their tails fully loaded and bristling, they advance on the enemy. As they get close, they point the conspicuous black-and-white tails over their backs and beat them up and down, so fanning gusts of their smell at the opposition.

Such proclamations as these send their warnings over long distances. Songsters may not even see one another. A landlord using smell may have left the area long before its message has been received by a rival. But such signals are also comparatively innocuous. Smells may be repugnant but they do not damage; songs may irritate but they do not cripple. So if an animal's need for somewhere to live is pressing, it may decide to ignore such warnings and move in anyway. The dispute, if it is to continue, must be conducted at somewhat closer range.

The contestants now demonstrate, and if possible exaggerate, just how big and powerful they are. Cats do so by arching their backs, bristling their fur and standing on tip-toe. Rats erect their hair and then turn sideways to one another so that their rivals will appreciate their full dimensions. Angry toads and chameleons inhale air and inflate themselves. Parrots erect their head-feathers and crests, and an angry elephant, advancing on its rival, nods vigorously and holds out its huge ears so that its head appears to have doubled in size.

These threats may be made even more explicit by giving particular prominence to the weapons that an animal will use if necessary. Crabs hold out their great pincers and wave them. Antelope and horses, which fight with their hooves, stamp at one another. Gulls attack with their beaks and blows from their wings so they lower their heads, point their beaks at the enemy and lift their wings slightly away from the body.

When most mammals fight they bite, so they threaten by drawing attention to their teeth. Dogs and cats expose them by snarling. An angry guinea pig grinds its big gnawing incisors together with a grating sound. A camel not only gnashes its teeth but produces a great quantity of saliva at the same time so that it, literally, foams at the mouth with rage. And a hippopotamus, when it rears up out of the water and opens its mouth in a stupendous yawn, is neither tired nor bored. On the contrary, it is displaying its massive peg-like tusks in the hope of terrifying its rivals.

Even such aggressive threats as these, made at comparatively close quarters, may not be sufficiently daunting to settle the issue. Sometimes that can only be done with physical contact. Yet even at this stage in the dispute, injury can be kept to a minimum if the contestants fight to mutually accepted conventions.

The strawberry arrow-poison frog of the Amazon rainforest, a jewel-like creature with a bright orange body and purple legs, follows rules that are entertainingly reminiscent of Japanese

Sumo wrestling. A male claims a territory about half a square metre and sits within it, usually on a buttress root thirty centimetres or so above the ground, proclaiming his ownership with cricket-like chirps. If a male without a territory challenges him, then the owner jumps down calling loudly. Sometimes that is enough to see off the intruder, but should he stand his ground, then the two square up and grab one another around the chest, pushing vigorously with their hind legs. Eventually one is likely to topple and then the other jumps on his back and pins him down with an arm-lock. As they wrestle, both continue to call loudly, as if each were protesting that he had no intention of submitting. When, somehow, the hold is broken, the grappling starts again and may continue for as much as half an hour as the pair hop, flip and roll over in the leaf litter. Eventually, one signals submission by creeping away, making only an occasional and somewhat muted call as if to maintain a little of his dignity, even in defeat.

Giraffes, when attacked by an enemy such as a lion, kick out with their hooves and are able to deliver extremely damaging blows. But when they quarrel among themselves, they restrict their actions to necking. Two quarrelling males stand alongside one another, usually shoulder to shoulder, but sometimes facing in opposite directions. They begin by craning their towering necks as high as they can. If neither is deterred by this, then one bends his neck outwards and swings it back so that the short blunt horns on his head strike the shoulders and neck of his opponent. His rival replies in the same way. As the battle gets more intense, the contestants straddle their fore-legs in order not to lose their balance. They would have little difficulty in landing kicks, standing at such close quarters, but they never do so. They are like heavyweight boxers who could cripple one another with a blow below the belt but refrain because that is against the rules.

Zebras conduct their contests to rules that are even more precise and complex. A zebra stallion is lord of a family group that includes several mares, who often attract the attention of other young males. If one approaches, the ruling stallion threatens the challenger by staring fixedly at him and wrinkling back his upper lip, so displaying the teeth with which he might, if provoked further, deliver a punishing bite. If the newcomer persists, then the tournament begins.

Round One. The opponents circle warily, each trying to bite the other's legs. As they get closer and become more persistent, one of them decides to protect his legs from these attacks by the simple expedient of sitting on them, and the round ends with both squatting alongside one another.

Round Two. The fighters now begin to shuffle around one another, still sitting on the ground, and try to snatch bites at their folded legs. The bout may end here with one of them scrambling to his feet and galloping off. But if neither of them gives up, then they move to the next stage.

Round Three. The technique now is neck-wrestling. One slams his neck across that of the other and bears down with such force that his forequarters are lifted off the ground. This continues for some time as the two fight in rising clouds of dust. Once again, the intruder has the opportunity to end matters and surrender. But if he does not, then there is no alternative but to enter the final punishing stage.

Round Four. No holds are barred. The two stallions scramble to their feet and attack one another viciously with both teeth and hooves. They rear up, striking out with their fore-legs, biting at one another's necks and legs, ripping and tugging with their teeth. Bunches of hair are pulled out of manes, ears get torn and hides so badly gashed that they bleed. This all-out battle must now continue until one decides that he has been punished enough. With a flourish, he breaks away and gallops off. The other may pursue him, but not far. Even though the loser is retreating, he can still deliver one final and extremely damaging blow, a kick in the face. Nor will the winner want to leave the mares that he has retained or just acquired, for they could be stolen while he is away by yet another stallion.

Restraint in battle is particularly necessary among those animals that have lethal hunting weapons. Rattlesnakes are armed with one of the most virulent of all venoms. Injected by a stab from their long curved fangs, it will kill a small rodent in seconds. These snakes too have their quarrels and in autumn, at the beginning of the breeding season, they fight among themselves. If they are not to kill one another, they must fight with great care. When two rival males approach, face to face, they put the sides of their necks together and rear upwards. As their back halves continue to advance, their front sections rise higher and higher. They sway sinuously from side to side, partly propping one another up. When their heads are up to a metre or so above the ground, one of them makes a final lurch upwards and falls heavily on his opponent, slamming him to the ground. They separate and then the wrestling starts again. The two may continue in this way for as long as half an hour, but at no stage does either attempt to bite his rival as he could very easily do.

So innocuous does this performance appear that it is often thought to be courtship between male and female and called a dance. And when the bout ends, it is not always clear to a human observer which is the winner. The consequence of the engagement, however, certainly seems to be that thereafter the two snakes keep out of one another's way. It is a paradox, but one that is not unknown in human affairs, that the most powerfully armed must necessarily in their quarrels be the most restrained.

Some species develop special weapons with which to fight with their peers. The fact that the males acquire such armaments only during the breeding season and that in many instances the females never develop them at all is convincing evidence that their primary function is not to defend their owners against enemies.

The biggest by far of all these temporary weapons are those grown by the bull Alaskan moose. The animal is itself a giant, the biggest of all deer, standing as much as two metres high at the shoulders. In the spring, a pair of bumps that are permanently present on his skull start to swell. Blood in the abundant vessels beneath the furry skin that covers them fuels the development of bone beneath. The bumps develop first into a column and then spread out into a huge plate. While they are growing, the moose takes great care not to knock them against anything, for a blow would not only be painful, but it could disfigure their final form. By August, their growth is finished and they have become the biggest antlers produced by a living animal. A large pair may have a spread of two metres. Now a ring of rough bone develops around the base of each one, blocking the supply of blood. The velvety skin that covers them splits and shrivels, exposing the white bone. For a short period, the velvet hangs from it in strips, but soon that is brushed off and the antlers are ready for action.

They are both offensive and defensive weapons. The broad central palm serves as a parrying shield, while the spikes around it can be used to slash hide, puncture a flank or even put out an eye. Contests begin with rivals walking obliquely towards one another. Each tips his head slowly from side to side making as plain as possible the full dimension of his antlers. And that may be enough. A full-grown bull in his prime may have as many as twelve spikes on each antler. A four-year-old, acquiring proper antlers for the first time, may only have six spikes on each of his. And although he will develop bigger and bigger versions each year, providing he gets enough to eat, he may not grow a really powerful set until he is about eight years old. So this initial inspection between rivals, measuring up one another, is of great importance and a young male, confronted with the spectacular armoury of a full-grown bull, may decide that the time has not yet come for him to make a bid for power.

If the two are evenly matched, however, they continue to approach until they are within a dozen metres of one another. There they pause, lower their heads and start to thrash the bushes in front of them with their antlers. And then suddenly, they charge. The great antlers clash together and interlock. As the animals push backwards and forwards, snorting with the effort, one of the spikes may break. If one contestant, tiring, disengages too quickly, he may be stabbed in the flank. If they both become exhausted equally, they may each be grateful for a respite and both withdraw. But within seconds they again lower their heads, thrash the bushes and charge. The battle may continue, round after round, for ten minutes or more before at last one concedes and gallops off.

These contests go on for about a month. By mid-September, most are over. Females have been won and lost. Old bulls have been defeated; younger more virile ones have increased the size of their harems. The bone at the base of the bulls' antlers, close to the skull, is partially reabsorbed. The weight of the antlers can no longer be supported and they fall.

That a bull moose should get rid of them at this stage is not surprising. Now that the fighting is finished he has no more use for them this season. Their huge spread impedes him as he moves through the forest and their great weight puts a considerable strain on his neck muscles. But four or five months later, he will have to start growing them all over again. It is more difficult to understand why such temporary armaments should have become so extravagantly large, for growing such gigantic weapons afresh each year must make huge demands on an animal's resources.

It seems that if a male battles with other males for the possession of a harem rather than a single female, his species will, inevitably, be embarking on an arms race. If he wins because his weapons are bigger than those of his peers, and his very victories prevent many of his rivals from breeding at all, then he will sire a greater number of calves than other males and pass on to them those genes responsible for his advantage. So, over generations, antlers become bigger and bigger, even though their cost, in terms of food necessary for their production, gets higher and higher. Nor may we suppose that the moose's armaments have yet reached their limit of extravagance. A now-extinct species of deer that lived in northern Europe had antlers with a spread of over three and a half metres and a weight one and a half times that of the biggest Alaskan moose.

Beetles have a similar tendency to develop horns. The Hercules, the rhinoceros, and, the biggest of all, the Goliath beetle which can grow to seven centimetres long, are among the large number of species carrying particularly spectacular weapons. They vary considerably in shape. Some are outgrowths of the head and curve backwards. Others sprout from farther back on the body, the thorax, and point forward. Some species have a horn of each kind, which meet above the individual's head like a pair of callipers. Beetle fighting technique does not involve charging or pushing. Instead, they endeavour to prise one another from their footholds and hurl one another aside. To do so, they use their horns as levers, forceps and bottle-openers.

The stag beetle's horns project directly forwards. The name 'stag' is not really appropriate, for these 'antlers' are not outgrowths of the head, like those of a deer, but greatly elongated mandibles so they are more accurately compared to the tusks of an elephant or even, since the beetle can open and close them, to the beak of a bird. Nor do fighting stag beetles interlock horns, as true stags do. Instead, when rivals meet, which usually happens on the branches of a tree, they grapple, trying to get a grip on one another with their huge forceps. When one succeeds, he heaves his opponent upwards, pulling his hooked feet away from the bark until all six are free. Then he turns and drops his rival to the ground.

One species of darkling beetle has mandibles that do not project forward at all but curve up and around his head so that the two together form a semi-circle. Like the stag beetle, he can move these mandibles, but looking at the insect away from his environment, it is difficult to see how they could be of any practical value, for there seems no way in which such 'antlers' could meet in a head-on jousting match. However, these beetles live and fight in burrows. When two males meet, the more aggressive of the two revolves within the burrow and then crawls towards his opponent as if he were going to pass over and above him, back to back. But as their heads draw level, he uses his mandibles to grapple downwards for the other's neck and if he gets hold, he will clamp on to it with such force that he draws blood.

Antelope and cattle, sheep and goats are armed with horns that are outgrowths of the skull covered with material similar to that which forms nails and hooves. Unlike the bony antlers of deer, these are not shed every year but retained throughout life, growing steadily. Ancestrally, they probably began as small forward-pointing spikes with which rivals struck one another on the flank as they fought side by side. As the animals evolved, these attacks became head-on and increasingly formalised. The cattle and the sheep families turned their contests into butting competitions. Ibex, in the mountains of Europe, have spectacular scimitar-shaped horns which they use in jousts with one another. The bighorn sheep of the Rocky Mountains have taken such duels to an extreme. Two rams charge at one another full-tilt, crashing together, head to head with the most spine-jarring shock imaginable and a bang that echoes for kilometres through the mountains. They have thick dense horns on their foreheads which curl handsomely around each side with which to deliver these batterings. They also have hugely thick skull bones, which must have evolved in parallel with the horns if the rams were not to split their heads during these battles.

The antelopes have ritualised their contests along different lines. Their horns have become long, gracefully curved grappling irons. Opponents lower their heads, interlock horns and then wrestle back and forth in a trial of strength. Many of these matches are conducted to rules that are even more elaborate than those followed by zebra.

Grant's gazelle, one of the commonest small antelope of the East African plains, has horns that are ringed with ridges and sweep elegantly backwards in a gentle curve. When two bucks compete with one another, they approach cautiously, holding their heads high with their ears pointing forward and their muzzles tilted up so that their horns slope back over their shoulders. They do not charge or even meet head-on. Instead, as they get close, each turns his head away to one side. When they are level with one another and their flanks are only a few centimetres apart, each faces forward again and nods, so that his horns are flaunted in full sight of his rival. When they are exactly alongside, they both stop and stretch their necks upwards with their heads turned towards one another so that they show off the prominent white patch they have on their throat and chin. Again and again they repeat this sequence of gestures. In more than half of these encounters, one of them will then decide that the contest is over and leave without a blow being struck.

But if this appraisal is not enough to convince one of them that his case is lost, he turns, lowers his head and interlocks horns with his rival. Now the duel becomes a trial of strength. Back and forth they push. If there is no obvious winner, as is usually the case, the two separate and move apart, grazing nonchalantly as they go. Only in one out of ten of these bouts is there a clear-cut outcome in which the weaker of the two acknowledges his defeat by suddenly disengaging his horns and running for it. Yet the mutual testing is enough to determine the social relationships between the bucks.

The antler flies of Australia settle their disputes in a similar way, although on a tiny scale. The males, who vary considerably in size, develop long antler-like outgrowths on their cheeks. If an ill-matched pair meet, the smaller retreats immediately. If, however, the two are of a similar size, then they place their antlers against one another and push vigorously. As they shove, so they rise higher and higher until their two hind pairs of legs are almost vertical and the front pair of each is lifted well clear of the ground. The smaller one eventually topples backwards and flies away uninjured.

Spectacular weapons and physical strength, however, are not the only factors that determine the outcome of battles, even those that have been ritualised to such a degree as this. An animal fighting on his home ground, whether he is an antelope or a bird, a frog or a fly, seems very often to have a psychological advantage over an intruder. One of the great pioneers of behavioural studies, Niko Tinbergen, demonstrated this with a classic experiment involving sticklebacks.

A male stickleback, when breeding time approaches, develops a bright red belly and chest and swims aggressively round the small nest he constructs in the weeds. If another male arrives, then he stands on his head, beats his fins in a vigorous threatening way and erects the spines on his back. There may be a skirmish in which the two fish attempt to bite one another and then usually the intruder accepts defeat. His red breast fades, he lowers his dorsal fin and slinks away as unobtrusively as he can. Tinbergen kept two pairs of sticklebacks in an aquarium which was sufficiently big for each to build his own nest and defend the territory around it. He then caught the two males and put each in a glass test-tube. When he put the tubes alongside one another in one territory, the owner, as expected, stood on his head and fanned his fins, and the intruder – also as expected – faded. The two test-tubes were then moved to the other territory. The erstwhile intruder, now back home, brightened immediately and recovered his confidence, whereas the other fish, in alien country, just as quickly lost his. Since both were enclosed in tubes, not a bite had been delivered nor even an aggressive current of water fanned in either direction. The only change in their physical circumstance was the visual appearance of their surroundings. That was enough to determine the outcome.

The ritualisation of conflict also extends to the ways in which disputes may be brought to an end. The vanquished concedes defeat with a signal which results in the victor stopping short of inflicting serious damage. The gestures that do this are simply the opposite of those that are used in threat. The strut is replaced by the cringe. Hair, that on an aggressive animal bristles, on a defeated one lies flat. A gull which proclaimed its threat by opening its beak and pointing it downwards towards its rival, now signals submission by shutting it and pointing it upwards. A chameleon which turned black with rage, in surrender turns white. And instead of shielding the vulnerable parts of its body as an animal does during a fight, the vanquished will reverse that action also and drop its guard, as though throwing itself on the mercy of its conqueror. A wolf rolls over and exposes its throat and belly to the teeth of its enemy; an agouti lies on its side, legs outstretched with its eyes closed; a wild male guinea pig, after screaming in fury, will turn his rump to his conqueror and suddenly, in as vivid a display of volunteered vulnerability as could be imagined, protrude his brightly-coloured scrotum. All these signals of surrender normally stop any further attacks. It is not only human beings who consider that kicking someone when they are down is against the rules.

Nonetheless, all these rules, conventions and rituals are not always completely effective. Battling male lions occasionally maul one another so severely that one dies of its wounds. Stags sometimes get their antlers so entangled that they cannot separate and they starve to death together. Elephants occasionally kill one another, and a vanquished rat, even though it may not be visibly injured in any way, after losing a fight may creep away and die.

But for the most part, fighting is a positive element in an animal's life. It ensures that the next generation of young have the most vigorous and powerful father. It results in individuals being spread out widely through the territory occupied by their species, so that maximum use is made of its resources. And, thanks to the rules of engagement that are understood and accepted by competing individuals, all this is brought about with only the minimum of injury and death.

NINE

Friends and Rivals

Many animals live in communities – herds and packs, flocks and swarms, gaggles of geese and troops of monkeys. The reason they do so varies. Gazelles on the open plains of Africa congregate in herds because it is safer that way. Lions live in prides and wolves in packs because their success in hunting depends on them working together. And pigeons may gather in flocks for no more complicated reason than that a great quantity of food is frequently to be had in a relatively small area.

But whatever advantage there may be, the practice creates problems. There will inevitably be squabbles over food, disputes over nesting places, arguments over mates. It is to the advantage of every individual that such disputes should be settled with the minimum expenditure of time and energy and that if there are fights, however ritualised, they should not need to be repeated at every subsequent encounter. In short, such communities work more efficiently if they have some form of social structure.

Some groupings are only temporary. Whooper swans, in autumn, fly down from the far north to take refuge from the bitter winter in the relatively milder climate of Britain. They assemble on stretches of water that they may use year after year. Each season, however, the group changes slightly. Some birds will have died and others come to maturity. So each year relationships have to be sorted out afresh. As two males try to collect the same piece of weed, they hiss at each other, stretching out their necks and quivering their wings. If they get really angry, they lunge and spar with great splashings and flappings. But just as sticklebacks become emboldened when they are close to their nests and intruders lack the confidence to oust a resident, so the social circumstances of the swan families largely determine the outcome. The bigger the family, the more likely it is to dominate. A male with a mate and several cygnets will win over one with only a single cygnet, who in turn will take precedence over a pair with no cygnets at all. And an unmated adult will defer to every other except those in the same social situation. Furthermore, cygnets, when they are feeding close to the rest of the family, take on the ranking of their parents and will see off another adult who is junior to them.

Great tits also assemble into temporary groups at certain times. During the easy days of summer they are dispersed throughout woodlands, but when the weather turns really cold, they gather wherever food is to be found. They do not have their families with them, like the swans, yet their disputes are settled equally quickly. Their seniorities have a quite different basis. A great tit has a black band running down its yellow breast. A bird with a narrow band will defer to one that has a broad band, just as a lance-corporal in the army with a single stripe on his sleeve will recognise the authority of a sergeant who has three. As a result, during this difficult period, the minimum time and energy is wasted in squabbles. Since senior birds cannot be everywhere, even juniors get their turn except in the harshest of circumstances.

The question that arises, of course, is how does a bird become a sergeant. The answer seems to be that the broad stripe develops on those birds that are well-fed as chicks and also on those that had parents with such broad stripes. So it appears that, to some degree, great tits have a hereditary aristocracy. Vigour and aggressiveness seem to be inherited along with the broad stripe, enabling the badges of seniority to be an economic way of settling matters at a time when it is important to conserve as much energy as possible.

Jackdaws remain in flocks throughout the year. Their seniorities are established early each season with a succession of fluttering disputes. But once they have been determined, the individuals concerned remember one another and the outcome of their last encounter. There is no need to spend more energy in repeating them. Instead, when two old rivals meet, one acknowledges its junior rank with the same signals of submission that it employed to end the previous conflict. It gives a quick nod of the head, exposing the vulnerable nape of its neck, and then turns away to allow the other to take the food it sought. These relationships spread through the entire flock and so produce a ranking order or hierarchy among the birds. The most senior individual is deferred to by all. Next there will be one who defers to the first but to no others, and so on throughout the whole community.

This kind of social organisation was first observed and described in 1922 among a flock of farmyard chickens. It was elucidated by carefully noting which bird pecked which and so became known as a pecking order. The term soon entered the general vocabulary since its relevance in human affairs was very quickly recognised. And while it is true that, in many ways, animal pecking orders are similar to rankings in, say, the army, the universities, the church or in boardrooms, it is also the case that neither the human nor the animal version is quite as simple or unchanging as it might seem to be at first sight. Jackdaws may have triangular relationships in which A is dominant to B and B is dominant to C, but when C meets A it turns out to be the winner. Sometimes there are alliances: B and C will unite to defeat A. And when the breeding season arrives, so do further complications. A young female jackdaw, taking up with a much more senior male will assume his rank – and exploit it to the full. Even so, once a jackdaw flock has settled down, the outcome of almost any encounter is largely predictable.

Caciques, black and yellow members of the oriole family that live in South America, have two separate hierarchies – one for males and one for females. It is only the females who build nests. They weave long club-shaped constructions that they suspend from the branches of a tree and they compete vigorously with one another for the best sites. They prefer to nest in trees standing on small islands which raiders such as snakes and monkeys find difficult to reach. They also habitually build close to the nests of large wasps, for these accept the presence of birds but are quick to defend their own nests against likely thieves and there-fore, coincidentally, the birds' nests that hang alongside them. As many as a hundred female caciques will build in the same tree. The scarcity of suitable sites necessarily brings them together, but it is also an advantage for them to nest in colonies since they can join together to mob and repel an intruder.

Nonetheless, there are raids. A wandering capuchin monkey may find its way there. A toucan can fly in and will take an egg or a chick if it gets a chance. So the best place for a female to build is in the centre of the colony as close as possible to the wasps' nest. Most robbers will have been seen off before they get to them. The females therefore fight with one another to possess such a site. They spar in mid-air with great ferocity, sometimes falling to the ground as they grapple with claws and beaks. The heavier bird of the two is usually victorious, but once a female has won a site she is given some degree of entitlement. As she grows older, she tends to lose a little weight, but newcomers who may be stronger and heavier than she will nonetheless defer to her.

The male caciques have their own hierarchy. They hang around the nest tree, displaying their golden-yellow rumps and flashing their blue eyes, sometimes in courtship towards a female and sometimes competitively at other males. They fight in the air and perch on branches facing one another, screaming aggressively. These confrontations usually produce a clear winner and in any further encounter between the contestants, the vanquished bird accepts the seniority of the other.

As soon as a female finishes building her nest and is on the verge of laying, the males compete to mate with her. The battle is vigorous and intense and often confused, for as many as a dozen males may be involved. After mating is over, the successful male will stay continuously within a metre or two of her for the next five days, by which time she will have completed laying her clutch. In this way he prevents another male from fertilising any of her eggs. After that he immediately looks for another mate.

If he discovers a female who has just mated with a junior male whom he defeated earlier in the season, he will drive him off and claim her.

Monkeys are so active, intelligent, inventive and inquisitive that when they live in large groups they develop particularly complex social structures. Olive baboons live on the savannahs of East Africa in communities that may be as much as a hundred strong. Their society is built around the females. The most senior of them is likely to be one of the oldest. Her daughters have a rank very close to her and are senior to the daughters of any other female. Subservience between them is signalled by a 'fear-grin', a kind of nervous smile, and by lifting the sole of the foot and pointing it backwards while at the same time raising the tail vertically.

Males are twice as big as the females and appear to be much more powerful, for they have huge canine teeth. But their position in the troop depends on their relationships with the females. A young male, either within the group or joining it from another group altogether, must form an attachment to a female before he is granted a permanent place. He achieves this by making ingratiating noises to his chosen mate, and by smacking his lips loudly when they meet. The relationship may take several months to achieve but he knows that he is succeeding when she allows him to groom her, which he does by combing through her hairy coat with his fingers and teeth, picking off flakes of skin and ticks with an assiduity and frequency somewhat beyond the limits of necessity. Once the partnership is established, he spends much of his time with her and her family, sleeping and feeding with them, and is accorded many of her privileges.

But he also has to justify his position by accepting challenges from other males. These confrontations have their own vocabulary of threat and submission. A threat is made by displaying the canine teeth in a huge yawn; a submission by a turn and a presentation of the rear. Confirmation of the outcome of an encounter may be made by the dominant male quickly grabbing his junior around the waist and momentarily mounting him in a gesture that, were it to take place between male and female, would be a preliminary to copulation.

The social order in the troop is made more complicated because a female may have more than one male friend. Two is quite usual. Not only that, but a male may also maintain close relationships with more than one female, though he usually restricts his attention to those within the same family. Human observers find very real difficulties in sorting out the ramifications of these relationships well enough to predict the outcome of any encounter, and if you watch a baboon troop you may be forgiven for thinking that the baboons themselves do not find it straightforward either. They constantly seek confirmation of one another's ranking and during a rest period or while they are feeding there is a continuous exchange of yawning threats, nervous smiles and submissive presentation of bottoms.

When the troop is on the move, each individual has its own position which is determined by its age and sex as well as its own rank. The senior females with their attendant mates and offspring travel in the safest place, the centre of the troop. The younger males travel alongside those females in their immediate families who have infants, ready to give them special help and protection. The youngest and most junior males, at the bottom of the hierarchy, act as scouts at the front of the advancing troop, looking out for danger from enemies such as a leopard that might be lurking in ambush. They face the greatest dangers. If the troop is attacked, they are likely to bear the brunt of it. If they lag behind the main party and get into trouble of some kind, none of the others will go back to help them. They are the young heroes whose valour may save the whole troop.

To what degree this role is imposed on them or whether they voluntarily, in an altruistic way, sacrifice themselves for the sake of the community is a much argued question. Parents, of course, regularly and voluntarily make sacrifices for their young. They provide them with food when it is scarce and when they themselves may be hungry. They regularly put their own lives at risk in order to protect their babes. This we do not find surprising, for by doing so they are achieving that major objective in their lives, the transmission of their genes to the next generation. But an animal that sacrifices itself altruistically for another quite unrelated individual would present great difficulty to accepted evolutionary theory. Such a tendency to self-sacrifice, if it appeared in an individual, would by its very nature be unlikely to be passed on to the next generation, for its owner will leave less progeny than those who act more selfishly. Therefore altruistic behaviour might be expected to quickly disappear during the course of natural selection.

Yet biology is full of exceptions. For example, there is one animal that does, on occasion, seem to behave in such an unselfish way. Unlikely though it may seem, this is the vampire bat.

A vampire bat feeds on blood and nothing else. It needs to drink at least half its body-weight every night. Collecting that is not easy. The bat has to land on a mammal, usually a horse or a cow, detect with its heat-sensitive nose just where there are blood vessels close to the surface and then shave away the skin with its triangular incisor teeth. Its saliva contains an anticoagulant that ensures that the wound will remain open long enough for the bat to complete its meal as well as an anaesthetic that reduces the likelihood of its victim being irritated by its attentions and shaking it off. It needs to be able to drink for about twenty minutes if it is to fill its stomach. To do all this takes luck as well as skill and a third of the immature bats in a colony may fail to feed at all on any one night. Even seven per cent of experienced adults will be unsuccessful. Yet if an individual does not get blood for two nights in succession, it will die.

Female bats live in small groups of about a dozen. For most of the year they have a young pup with them. They not only provide it with milk, but when they return from a successful night's raid, they regurgitate blood for it. But a bat with a full stomach will also give blood to another adult in the roost who has failed to feed. The recipient may be a relative – a sister, a daughter or a mother – in which case the behaviour is explicable in evolutionary terms for such relatives share a very high proportion of their genes with one another. The action therefore aids the propagation of the genes in the same way, though to a lesser degree, as does that of a parent caring for its young. But careful investigations using genetic fingerprinting techniques have shown that those being fed are sometimes not closely related at all. Is this, then, an example of totally unselfish behaviour or could there be an advantage to an individual that acts in this way?

It would be to the bat's benefit to give blood to a starving companion if she could be sure that when she herself had bad luck and failed to feed, the recipient could be relied upon to behave in the same way. And that proves to be exactly what happens. Although the dozen or so bats in a cluster move frequently from one roost to another, they usually do so as a group and one individual will habitually hang close or even next to a regular companion. These particular individuals might not unreasonably be described as friends for they groom one another lavishly and recognise one another by the characteristic sound of their voices. A cheat who solicited blood and took it but did not repay the debt when required would soon be detected. So the giving of blood between these unrelated friends is a reciprocal form of altruism that benefits both partners. It is thus a characteristic that can be selected and encouraged by the processes of evolution.

Altruism between relatives is much commoner. The dozen or so dwarf mongooses that make their home in a termite hill are nearly all related. Most are the children of the breeding couple who originally founded the group, but while these two are still alive and producing babies, no others in the group will breed. Instead these juniors will devote their time to looking after the young and helping with the welfare and safety of the group as a whole.

Some, usually young males, act as sentinels. When the family moves out, searching for beetles and other insects, they climb to the tops of bushes and termite hills, scanning the surrounding countryside and the skies for enemies. Their greatest danger comes from hawks. If they see one, they whistle a warning and everyone takes cover, sprinting down their tunnels in the termite hill if they are near enough or hiding beneath bushes. When danger is past and the group resumes foraging, others will take over guard duty so that everyone gets a chance to feed.

Babies that are too small to venture out on these expeditions remain back at the termite hill. There they are looked after by other members of the group, usually young females. They groom and play with their charges, retrieving any that wander too far from the entrance holes. They catch insects for them. Even more remarkably, young females who have never been pregnant produce milk from their breasts and suckle them, a very unusual phenomenon.

The dwarf mongooses are at such risk from predators that were they not to work in teams, they would find it very difficult to deal with the simultaneous demands of protecting themselves when out foraging and of feeding and caring for the young. That the adult children of the founding pair should help in this way is understandable, since by doing so they are helping to propagate their own genes. But some of the group are unrelated immigrants from other families. Why should they help? It may be because by doing so they are able to feed more safely in such a guarded group; or that they are older than any of the other helpers of the same sex and therefore are the ones most likely to take over the colony and breed themselves if and when the founders die. If that happens, then their future children will be looked after by the young that they themselves are helping to rear.

One mammal has taken this division of labour within the family to quite extraordinary lengths. It is one that is normally never seen in the wild for it spends its entire life underground in long tunnels that, with their loops and branches may extend for over two kilometres. The only signs of its existence above ground are craters of earth thirty centimetres or so across from which occasionally spurts of fresh soil erupt. These are the spoil heaps of naked mole-rats.

These animals are about the size of a small brown rat, although they seem somewhat smaller because they totally lack fur. Their skin is pink, wrinkled and baggy. They have two pairs of enormously long incisor teeth which protrude like pincers so far in front of the jaw that their lips close behind them, so allowing the animal to bite through the ground while digging its tunnels without getting any earth into its mouth. Their eyes are tiny and in any case are of little use to their owners for only when they are near the exit craters can they glimpse any light and most of their lives are spent in total darkness. They have long bony tails which they use like antennae, wagging them vigorously from side to side when, as they must do in such narrow tunnels, they run backwards.

The country in which they live is very arid for much of the year and as is usual in such areas, many of the plants that grow there store water in underground tubers. Some are as big as footballs and it is these that provide the mole-rats with most of their food. However, the plants grow in a very scattered and random fashion. As the mole-rats have no way of discovering where they are, they may have to dig many metres of tunnel before, by luck, they strike one. Then, however, they exploit it with care. They first cut a hole in its rind and then hollow out the centre pulp leaving the rest of the outer skin intact. While this robs the plant of much of its food store, it does not kill it and the rats will sometimes block off the tunnel leading to a severely plundered plant to allow it to recover. Then they return to it later.

The members of the mole-rat community vary considerably in size and perform a range of behaviours. Scientists are still unsure about the links between size, age and behaviour, but some things are clear. The smallest animals, about eight centimetres long, work very energetically digging new tunnels and cleaning the old ones. As soil is excavated in a new tunnel, so a line of these workers, labouring as a team, kicks the loosened earth backwards until it reaches one of the exits. The last in the line is responsible for kicking it out, so building the crater surrounding the hole. Some workers, often slightly bigger, spend most of their time lounging in the central nest chamber; others again are involved in defending the colony. Although the workers block off the exits except when soil is being thrown out, sometimes a snake does manage to slither into the tunnels. While most of the mole-rats squeak in fear and scuttle down to the nest chamber to take refuge, some will rush to meet the enemy and attack it fearlessly, biting it so severely with their pincer teeth that they are quite capable of killing it.

All these groups contain both males and females. But the sexuality of nearly all of them is quiescent. Only one female in the community is sexually active. She is the queen. She may live for thirteen years and she can produce a litter of young every eleven weeks. Each litter contains about a dozen pups. Her mates are one to three males drawn from the ranks of the soldiers. Their accession to sexual potency, however, does not last for long. The effort involved seems to wear them out and after a year or so of activity, they die and are replaced by others from the same group.

The queen does not dig, nor does she involve herself in the defence of the colony. Once pups are weaned, she hands them over to the half-workers who are, of course, the elder brothers and sisters of the infants and who feed them on fragments of chewed-up tubers. Apart from giving birth and suckling, which she does in the nest chamber, she spends most of her time patrolling the tunnels and harrying her subjects, nipping some, particularly the females, and nuzzling others. When eventually she dies, she will be replaced by one of the half-workers.

How does what looks like a despotic empire survive over evolutionary time? Why do the workers not revolt? There are two theories, which for the moment remain unresolved. It may be that the queen is actively suppressing the workers by inhibiting their reproduction, either physically or chemically, or it may be that she provides the workers with a signal that she is there and reproducing, giving them the information they need to inhibit their ovaries. The second interpretation has the advantage of suggesting that the whole system benefits the workers as well as the queen – they voluntarily go along with things because it somehow increases their fitness. Furthermore, if inhibition were active, we would expect there to be signs of an arms race, and for some mole-rats to have abandoned the social life-style completely as the different interests of queens and workers played out. Cape mole-rats do indeed live a virtually isolated life, but it is not clear if their life-style represents the loss of sociality, or a separate lineage. In the end, the question of how social mole-rats organise their societies will be resolved by more careful experimentation, not simply by theoretical musings.

What might be the advantage to a worker of being a member of this community when the cost is the loss of its ability to reproduce directly? First, it is able to live in areas that would otherwise be closed to it, for a single animal or even a pair could hardly dig the long tunnels necessary to locate the great tubers. (Cape mole-rats live in far more welcoming environments.) Second, even though it is sterile, it nonetheless is assisting in the propagation of its genes, for some of the infants in the colony's nurseries are its full brothers and sisters and on average each of these, therefore, carries at least half its genes. On average that is the same proportion as its own babies would have if it mated with an individual from outside the community.

We do not know what evolutionary pathways were travelled by ancestral rodents that led to the development of this bizarre society nor the exact circumstances that made it necessary or desirable. Whatever they were, they have occurred several times in the history of life. Rodents, being mammals, are among the last animal groups to appear on earth and certainly the most recent to have started down this route. Insects have a much longer history. They were already abundant two hundred and fifty million years ago, and several groups of them began to live in large communities. Members of the cockroach family were the first of them to do so. They were the ancestors of the termites of today. By one hundred million years ago, the closely related families of wasps and ants were flourishing and each of them also, independently, evolved this way of living.

The leaf-cutter ant of South America has one of the most complex of these societies. From several important and significant points of view, leaf-cutters and mole-rats lead similar lives. First, both live underground and are thus able to defend their colonies. Second, both have concentrated the responsibility for reproduction on to one individual, a risky thing for a community to do under most circumstances but one that becomes much less so in the near-absolute security of an underground fortress. And third, both have reproductive females who are very long-lived so that their adult sterile children are able to assist in raising later generations of infants.

But whereas the mole-rat workers may move from one role to another, the ant workers are irrevocably fixed in their castes, and whereas the mole-rat community contains several dozen individuals, a leaf-cutter colony contains several million.

The leaf-cutters solve the difficult problem of digesting the cellulose which forms such a large part of plant tissues by cultivating a fungus to do it for them – as many termites, including the bellicose species, do in Africa. The leaf-cutters in America, however, invented the technique for themselves and use not dead vegetation but leaves and stems cut from living plants.

The nest of a leaf-cutter colony is gigantic. On the surface it appears to be a low mound of largely bare earth, rising in the centre to about a metre above its surroundings, but below ground it descends as much as six metres in a maze of chambers and passageways. Well-worn paths several centimetres wide lead from some of the thousand or so holes that stud the surface of the mound and extend into the surrounding forest for as much as a hundred metres. Along them, day and night, march continuous processions of ants, sometimes ten abreast, most carrying a small freshly-cut segment of leaf which they hold in their jaws hoisted above their backs like tiny flags.

When they arrive inside the nest, they drop their loads on the floor of a chamber and hustle out again, back to the cropping site, following a scent trail laid down initially by the scouts who found the tree and now reinforced a million times by their fellows.

The abandoned leaf fragments within the nest are picked up by a different caste of worker. Whereas the porters were about the size of house flies with heads 2.2 millimetres across, these are somewhat smaller with heads measuring only 1.6 millimetres. They lick the leaf segments to remove any spores or bacteria they might hold which would contaminate the cultures within the nest, and cut them up still further. A caste of even smaller workers now takes over, champing the fragments into a moist pulp and adding little droplets of anal fluid that will help to break down the leaf tissues chemically. They carry the result to special garden chambers. Each of these contains a ball of grey spongy material ranging in size from an orange to a melon. These are the fungus gardens. The ants carefully insert the processed leaf fragments into holes with which these masses are honeycombed. Here the tiniest and most numerous workers of all take over. Their heads are only 0.6 millimetres across. Only these dwarfs are small enough to crawl inside the spongy gardens. They clamber over the leaves and with their delicate forceps-like jaws pluck tufts of fungal threads and plant them on the macerated leaf surface. The fungus grows very quickly, completely covering the leaf fragment with a tissue of white threads within twenty-four hours. The dwarfs tend these gardens with great care, weeding out alien fungal threads that may have got in. As the fungus matures, the ends of its threads swell into tiny knobs. These are harvested by workers of all castes. Some they eat themselves. Others they take away and feed to the grubs that are kept in the nursery chambers.

The community is guarded by yet another caste, the soldiers. They weigh three hundred times as much as the dwarfs and are as big as bees. Their heads are swollen to accommodate the muscles of their huge jaws with which they can chop a marauding ant in two and slice a painful cut in a human being's hand.

At the centre of this great nest sits the queen. Compared with all her progeny, she is gigantic, as big as a cockroach and her entire life, once she has established the colony, is restricted to laying eggs. When she took off on her nuptial flight, several males may have mated with her while she was on the wing. Their sperm entered a special pouch in her body where it can remain alive and viable throughout her life. However, as with many insects, her unfertilised eggs can also develop and sometimes she produces these by shutting off the passage that leads from the sperm pouch to her oviducts. Before she dies, she may lay as many as twenty million eggs, both fertilised and unfertilised.

She is tended by bevies of medium-sized workers who collect the eggs as they emerge from her body and carry them away to the nursery chambers. The unfertilised ones become males, which have wings. When they mature, they will fly from the colony without having contributed anything to it – their role is merely reproductive. They are in effect flying sperm. The majority of the queen's output, however, are fertilised and these become females. Their caste depends on the way they are treated by their nurse-maids and the food they are given. A tiny proportion of them become new fertile queens and they, too, in the appropriate season, will leave the nest and fly away with the males to take their chance in establishing new communities. But the vast majority remain sterile and will reinforce the ranks of soldiers and porters, leaf-processors and dwarfs who are all needed if the colony is to thrive.

The ancestral ant, when it first evolved from a solitary wasp so many tens of million years ago, lived in small groups of cooperative breeders. The lives of modern solitary wasps (there are no solitary ants) are short. Each has, by itself, to cope with the essential tasks of life, collecting food, avoiding enemies, and reproducing. Failure in any one of these labours results in the extermination of the individual and the genes it carries. Large colonies of ants, such as the leaf-cutters, multiply their numbers and allocate different jobs to different castes, keeping all these activities going simultaneously. Any failure in any aspect is likely to be only temporary and can quickly be put right without endangering the colony's survival. At one and the same time, leaves are being gathered and gardens cultivated, eggs are being laid, defences maintained and new generations reared.

Viewed from one perspective, it might seem that here altruism and self-sacrifice have reached their apogee with many millions of individuals foregoing their full development in order to serve just one, their queen. Some scientists have even argued that the ant colony can be considered not as a community but as something like a single individual, a super-organism whose members are genetically very similar. From this point of view, the super-organism lives not for just a year or two as a single ant might do, but for several decades. It is so secure that only its specialised tentacles stretch out into the forest and the vast bulk of it never appears above ground at all. And it exploits the plants around it with unrivalled efficiency and a relentless persistence. It can strip a bush of all its leaves in a single night and collect flower petals from the topmost branches of a tree. It is not the leaf-eating monkeys, nor the fruit-eating birds, neither the battalions of caterpillars nor indeed the human beings trying to grow crops within the leaf-cutter's reach, who are the masters of the land where it lives. It is the tiny ant that, by multiplying and organising itself into a society, has become a monster.

TEN

Talking to Strangers

The raised eyebrow, the curled lip and the exchange of hormone-laden spittle are very effective ways of communicating at close quarters, but animals also need to send long-distance messages to strangers of their own kind living in other communities. They even on occasion have to converse with individuals of another species. For both purposes, they have to use rather different techniques.

The message most frequently declared by one species to another is simple and straightforward: 'Go Away!' An angry elephant charging with outstretched ears, trumpeting as it comes, makes its meaning perfectly clear very quickly. Other animals of all kinds are in no doubt about it, whether they are prowling lions or over-confident photographers. But inter-specific communications can be rather more complicated than that.

The honey-guide is a lark-sized bird that lives in east Africa. Its diet is insects of all kinds and it has a particular penchant for the grubs of honeybees. Getting them, however, is not easy. Wild African bees build their nests in hollow trees or clefts in rocks. The honey-guide's beak is slender and delicate so the bird cannot cut away wood, still less chip stone. If it is to get its favourite food, it has to recruit a helper. Usually, that is a human.

In northern Kenya, where honey-guides still live in some numbers, men of the semi-nomadic Borana people specialise in collecting honey. When one sets out to do so, he begins by walking into the bush and whistling in a very penetrating way, blowing across a snail shell, a seed with a hole in it, or just using his clasped fists. If he is within the territory of a honey-guide, the bird will appear within minutes, singing a special chattering call that it makes on no other occasion. As soon as the two have registered one another's presence, the bird flies off with a peculiar low swooping flight, spreading its tail widely as it goes so that the white feathers on either side of it are clearly displayed. The man follows, whistling and shouting to reassure the bird that he understands its summons and is following.

The bird may now disappear for several minutes. When it comes back, it perches high some distance away, calling loudly and waiting for the man to catch up with it. As the two travel together through the bush, the bird stops and calls more frequently and takes lower and lower perches until, after maybe a quarter of an hour, its song changes into one that is low and less agitated. Having repeated this two or three times, it falls silent and flutters to a perch where it stays. Beside it will be the entrance to a bees' nest.

It is now up to the man to take the initiative. If the day is hot, a stream of bees may be buzzing in and out of the entrance.

Something has got to be done to pacify them if the man and the bird are not both to get badly stung. The man lights a fire close to the nest and, if possible, pushes burning sticks into holes beneath it so that smoke swirls up around the nest itself. With the bees partially stupefied, he now opens up the tree with his bush knife or pokes out the nest from a rock cleft with a stick and extracts the combs, dripping with rich deep-brown honey. That is what *he* wants. But not the bird. It prefers grubs. Even so, tradition demands that the man should leave at least part of the honeycomb for the bird, spiked on a twig or put in some other prominent place. The honey-guide can now get its share. It flies to the remains of the wrecked nest and pulls out the fat white bee-grubs from the cells of the combs. It also, very remarkably, feeds enthusiastically on the wax. It is one of the very few animals that can digest it.

The bird does not find its bees' nests by accident. It has a detailed knowledge of its territory and knows the exact location of every bee colony within it. Watchers in camouflaged hides have observed a bird visiting every one of its bees' nests, day after day, as though checking on their condition. On a cold day, when the bees are quiescent, it may hop on to the lip of the entrance and peer inquisitively inside. When the bird starts guiding the man, it does not wander about at random but leads him directly to the nearest nest. And the reason it leaves him for a short period just after their initial meeting is because it makes a quick flight to the nest it has in mind, perhaps to check that it is still flourishing. Furthermore, if having reached the nest the man for some reason does not open it, the bird, after a pause, will once more give its 'follow-me' call and lead him to another.

The fact that the bird has a digestive system specially adapted to dealing with wax suggests that it has been eating from bees' nests for a very long time and that therefore the relationship with man is an ancient one. Human beings have certainly been collecting honey in this part of the world for some twenty thousand years as is proved by rock paintings in the central Sahara and in Zimbabwe that show them doing so. The honeyguides themselves evolved around three million years ago, and may have begun this unique relationship with our ancestors shortly thereafter.

Messages exchanged between different species are not always sent to friends. Enemies also find it profitable to communicate. Thomson's gazelles, like springbok and several other small antelope, when chased by hunting dogs or hyaenas, behave in a strange and dramatic way. They stot, leaping into the air with their legs held stiffly straight downwards and landing on all four feet at the same time. Often they make several of these leaps one after the other, so that the animal appears to be bouncing across the ground. It seems very odd that, at a time when its life is in real peril, when its overwhelming priority must be to flee, it should find the time and the energy to indulge in apparently gratuitous gymnastics.

If gazelles are chased by a cheetah, they rarely behave in this way. The cheetah usually starts its hunt with a stalk, approaching closer and closer to the herd and apparently concentrating on one particular animal grazing in an approachable position. When the cheetah gets within range, it breaks cover and bursts into a sprint. Once it has started the chase, it concentrates on its target. It may be distracted by other gazelles running nearby or even crossing in front of it, but it seldom selects another victim, and either overhauls it within a hundred metres or so and pounces, or gives up. In such chases as that, the gazelle generally does not pause to stot. It runs for dear life.

Dogs, however, hunt in a quite different manner. They work in packs. Once they find the gazelles, they pursue the herd relentlessly. They show no sign at this stage of having selected one particular victim and often harry different individuals in turn. It is at this stage that the gazelles start stotting. As well as potentially warning its fellows of the presence of the predator, each may be sending a message about its fitness and therefore its ability to outrun the dogs. The higher and more frequently it stots, the stronger it must be. The animal upon which the dogs finally focus their attention will be one of those whose leaps were relatively feeble. The high-stotters escape without having had their fitness put to the test.

But if an individual gazelle is so strong that it really can outrun the hunting dogs, why should it spend time and energy in stotting just to say so? The answer may be that by doing this it avoids an even greater expenditure of energy in a long chase, that it does not risk injuring itself in collisions or falls, and that if it is threatened by a second hunter immediately after the first, it will not already be exhausted and vulnerable.

The message of the stot to the predator is, necessarily, a truthful one. A gazelle cannot appear to be fitter than it is. A plover, on the other hand, tells lies to its enemies. If, during the breeding season, you approach a ringed plover incubating its eggs on a pebbly English beach, it will sit tight, relying on its excellent camouflage for concealment until you are within a few metres of it. Then, when discovery seems inevitable, it will suddenly run off, shrieking loudly, with one wing raised above its back, the other held low and trailing as though it were crippled and unable to fly. If you were hunting for food you might well follow it, thinking that such an injured bird would be easily caught. You would be wrong. After some metres, as you get close to the bird, it opens both wings and effortlessly takes to the air. And you have had your attention distracted from its eggs, which still lie safely in their scrape.

It is easy enough to describe what a human being might be thinking during such an encounter. It is impossible to know what is going through the mind of a plover. It is certainly not warranted to suppose that it is consciously sending a signal saying 'I'm injured – chase me!'. Its behaviour may be no more than an elaboration of the emotional conflict that comes when a threatened bird is torn between the desire to flee and the urge to stand its ground and protect its nest. In such a situation, it may run with wings half-extended, neither staying nor flying. The action, nonetheless, may succeed in diverting the attention of an intruder away from the nest. Those birds that behave in this way will therefore have greater success in raising chicks. So the response becomes enhanced and genetically implanted in the plover's behaviour patterns.

Animals must also, on occasion, send long-distance communications to others of their own species. They may need to warn them of danger or to invite strangers from other groups to join them as mates. Out in open country, where gazelles graze and plovers fly, such signals can be visual. Pronghorn antelope live on the hills and prairies of the American west. If a coyote or some other enemy approaches, they broadcast warnings to each other with their buttocks. Each haunch carries a patch of white hair underlain by a disc of muscle. If the animal senses danger, it contracts the muscle so causing the long hairs to spread out into a huge rosette which reflects light with particular efficiency.

When the muscle relaxes, the disc contracts immediately. So bright white flashes are produced that are visible two kilometres away. Other pronghorns, seeing them, repeat them, so that the warning of danger spreads quickly across the country as though it were being sent by heliograph.

The pronghorns reinforce these visual warnings with a powerful scent emitted from glands that lie within the rosettes. Smell, however, is not a very efficient medium for long-distance signals. It will only travel in the direction that the wind happens to be blowing and in still air it will scarcely move at all. Even so, many moths use it very effectively when summoning a mate. A female emperor moth who is ready to breed, discharges a perfume from glands between the segments of her abdomen. The male has long feathery antennae with which he can detect and identify it even if it is so diluted that only a few molecules reach him. Once stimulated, he moves steadily upwind, zig-zagging back and forth through the odour plume, until he reaches the source of the smell. In favourable conditions, a female may attract males from a kilometre or more away.

Long-distance communication is particularly difficult in thick forest. Smell is even less satisfactory than it is in open country for there is seldom even the faintest breeze. Visual signals cannot be seen more than a few metres away. Here the most efficient medium is sound.

But all sounds are not equally effective. High-pitched calls are quickly doused by thickets of trunks and leaves. Complex trills and twitters, so often sung by birds elsewhere, are almost indecipherable at a distance, for the surfaces of leaves reflect sound and cause multiple echoes. So birds that live in thick forest, particularly those that have large territories and therefore need to be heard at a distance, tend to have calls that are not only loud but low in pitch and simple in structure – members of the same species living in different environments may sing slightly different songs. In other words, birds can have accents. The names of many birds are based on their calls, so it is no surprise that the deep forest is the home of the mot-mot, the potoo and the screaming pi-ha. The South American three-wattled bell-bird makes one of the most penetrating of all calls. Sitting in the topmost branches of one of its singing trees, it swells its throat and expels an extraordinarily piercing two-note metallic call; and it repeats it again and again with a persistence and monotony that can become maddening to a human traveller.

Spotted woodpeckers send simple low-pitched long-distance messages by drumming with their beaks on dead resonant branches. The action is an adaptation of the one they use when chiselling out their food. The huge black palm cockatoo that lives in the forests of northern Australia supplements its vocal screeches and deep whistles with a similar mechanical noise, but it has invented a truly novel way of doing so. Breeding pairs signal their ownership of a territory by breaking off a small stick with their claws and beating it against a hollow tree. They are the only birds known to use a tool to make sounds.

In the densest foliage, pairs of birds even a few metres from one another may have difficulty in keeping in visual contact. Sound, once again, is the most efficient way of doing so. Since the calls of a mated pair in their territory do not have to travel far, they can be higher-pitched and more complex than the long-distance signals. The bou-bou shrike, for example, lives throughout tropical Africa and frequents thick cover, whether it be forest or shrubbery. Its song is clear and flute-like with long melodic phrases. In fact, this is a duet, a phrase from one bird being followed without an audible break by a second from its mate. Sometimes one bird will do no more than insert a single note in an otherwise continuous sequence sung by the other. So well do these separate contributions fit together that it is almost impossible to detect that there are two singers unless you are so placed that you can see them both simultaneously. Indeed, the prevalence of duetting was not immediately recognised. Now, armed with devices to record calls and sonograms to analyse them, ornithologists are discovering that a great number of birds use this way of keeping in communication with one another. In two square kilometres of South American rainforest there may be as many as a dozen different species of birds that do so. It is rather touching to discover that duetting pairs, as a rule, remain together season after season, if not for life. Having developed the technique, they also practise it as a way of reinforcing the bond between them, singing their complex duets even while sitting next to one another on a branch; and sometimes, if one of the pair is absent, the lonely bird will sing the full elaborate melody, filling in the missing parts itself.

The forests and woodlands of Europe are not as dense as those of the tropics and this may be the reason why duetting of such precision does not occur here. But in the darkness of night the problem remains, and owls, one of the very few nocturnal birds, do indeed duet in a relatively simple way. The famous *to-whit to-whoo* of the tawny owl is actually the call of two birds. The first sings *to-whit* and before the last note has faded, the second bird answers *oo-oo*, so that the two calls appear to be one. These checks on one another's position are not always exchanged between male and female. Sometimes two tawny owls may be answering one another, establishing between them the frontiers of their hunting territories.

The same kind of duet between rivals is also performed in the darkness of underground tunnels by the blind mole-rat. Unlike its near relative the naked mole-rat of arid areas, this animal is fully-furred. It looks like an animated hairy sausage. It has no visible eyes, no ears and no tail. Apart from four stumpy legs, its only external features are the two pairs of long curved incisor teeth at one end which are its digging machinery. It lives in comparatively fertile areas, excavating long tunnels in order to find its food of roots and tubers, and instead of forming communities like its naked cousin, it is a confirmed solitary. If one should meet another outside the breeding season, the two may fight to the death. They manage to keep out of one another's way by signalling their positions. One will bang its head on the roof of its tunnel, beating it rapidly four times in the space of half a second and repeating the call eight or nine times in succession. Having drummed, it then presses its jawbone against the side of the tunnel to feel for the vibrations of an answering tattoo. If it detects one, it will reply and the exchange may continue for some time until, with their relative positions established, they return to their solitary tunnelling. In their subterranean corridors, neither sight nor smell nor sound could provide such information. Only seismic vibrations will do so.

Darkness compels many nocturnal animals to make their communications with sound. Geckos, the little lizards of the night, repeat simple musical notes. Insects shrill, dogs howl and lions roar; and during the breeding season, swamps full of frogs can be so noisy with choruses of yelps and moans, clicks and burps, that it is difficult to hear oneself speak. The babel may, however, be more confusing to us than to the frogs, for many species have ears that are sensitive only to the particular frequency band in which they themselves sing. And if two species do happen to sing around the same pitch, they will carefully space their calls, one waiting for the other to pause before it slots in its own signals.

Like all mating messages, those sung by frogs are characteristic of their species, proclaiming just what kind of animal they are and therefore what kind of mate they are seeking. But some give more information than that to prospective females.

The male mud-puddle frog of Central America, singing by himself and without competition, makes a relatively quiet mewing *aow-aow-aow*. This is the unique call-sign of his species and alerts female mud-puddles of his presence. But if he is in company, as is often the case, he has to do more to secure the attention of a female. He must add a second phrase, *chuck-chuck-chuck*. This gives the female a hint of his particular character. The bigger he is, the lower-pitched he is able to make this second phrase. Females prefer big strong males and so choose a bass rather than a tenor. Since that is the case, why do not big frogs always give their full call? Because their songs are heard not only by their mates but by their predators, fringe-lipped bats.

The bats find the frogs by listening for them. The frogs are well aware of this, for the silhouette of a bat flying across the moonlit sky is enough to silence an entire pond. Furthermore they apparently realise that the two elements in their call are not equally risky. The first part – the *aow* – is soft and gentle and therefore difficult to locate. The *chuck*, however, with its hard-edged staccato notes, is much easier to place and therefore much more dangerous to utter. So if a male mud-puddle is by himself and the only one calling, any female within range will come to him and there is no need to risk a *chuck*. But if the pond is full, he has a more difficult decision to make. To get a female, he has to declare his superiority over others. And that boast, every night of the breeding season, is the downfall of hundreds of braggart mud-puddles.

There is one other way of communicating in darkness. Instead of signalling with sound, you can use light. The cold glow produced by luminous organisms has an eerie quality to us, since nearly all the light we create and experience is inextricably connected with heat. Animals, however, generate their light chemically. A complex protein called luciferin, when mixed with small quantities of an enzyme called luciferase, will combine with oxygen and in the process give off a bright glow. Each luminescing organism has its own particular version of these compounds, but the reaction is a purely chemical one and it can be replicated in the laboratory with synthesised ingredients. Indeed, the dried bodies of small marine crustaceans that luminesce in life will glow brightly if they are simply moistened with water.

The commonest terrestrial luminaries are fireflies and glow-worms. They are neither flies nor worms but night-flying beetles, and they carry their light-producing chemicals in the hinder parts of their abdomen. Originally, they may have used their tiny torches simply to enable them to see where they were going. Many still do so. A female North American firefly, as she comes in to land, flashes her tail-lamp with increasing frequency until, just before touch-down, her flashes fuse into a continuous glow. As soon as she comes to a halt, she switches off entirely.

But these flashes are also used for signalling. The system is not unlike the Morse code as once used by naval signallers when working with hand-lamps, except that the fireflies' version is very much more complicated. Morse uses just two kinds of flash, one short and one long. Firefly flashes vary considerably in length, some lasting for as much as five seconds and others being repeated forty times within a single second at a speed so swift that our own eyes are unable to perceive the intervals between them. They also vary in the rate at which they are transmitted, the number of flashes in a sequence, the time between the signals and the pitches of intensity within the flash. All these variations have their significance.

World-wide, there are at least a hundred and thirty different species of firefly. The most spectacular are those that live in the mangrove swamps of South-east Asia. A journey in a canoe through certain creeks in Borneo and Malaysia at dusk may be uncomfortable because of persistent attacks from clouds of mosquitoes, but the reward is one of the most magical of all natural spectacles. As the sun sinks and dusk falls, scattered flashes begin to blink in the mangrove trees. Arcs of staccato lights, like the plumes of tiny fireworks, curve through the gloom as single beetles fly from one branch to another. Minute by minute, their numbers increase. Branches silhouetted against the sky become laced with tiny points of flashing green lights until the whole tree seems to sparkle and flicker. That sight itself is enough to take your breath away. But an even greater astonishment is to come.

Slowly the confusion of flashes begins to resolve itself into an order as the many thousands of insects synchronise their rhythms. Eventually, the whole tree, as a single unit, begins to pulse with light. Not all the mangrove trees are suitable for these displays. Many are inhabited by tree ants which will seize and kill any beetle that lands among them. So those trees that are illuminated often stand isolated above their rippling reflections with their pulsating lights outshining the stars that glitter in the black velvet sky behind them.

The speed of their pulsing is so fast that it would not be possible for the insects to take their cue visually from one another. They must each have an internal metronome that beats with such accuracy and so identically that once they are locked together, the entire multi-thousand assembly switches on and off in perfect unison. Furthermore, the pulse-rate varies slightly with the temperature. The colder the evening, the slower the rhythm. Yet all their metronomes vary to exactly the same degree so the synchrony is maintained.

Only the males display. They throng the trees in such numbers that were they not to synchronise, their signals would become confused and unrecognisable. Furthermore by collaborating, their throbbing rhythms, reflected in the water, are visible half a kilometre away across the swamp so that females searching for their own species can identify them from some considerable distance away. As they land on the branches, they mate with the first males into whom they blunder, while the less lucky males around them optimistically continue with their displays.

In the eastern United States, fireflies operate individually. There are several different species often occupying the same meadow or woodland. All the males are much smaller than the females and outnumber them by as many as fifty to one. Finding an unmated female is an extremely competitive business for a male firefly. The displays start soon after sunset and last for an hour or so. The females emerge from their burrows and take up their positions while the males cruise through the air flashing their enquiries, each species using its own characteristic code.

One produces long bursts at the rate of a flash every half second. Another makes two flashes a second apart and then stops and waits. A third makes only single sporadic flashes. The commonest response from the females of all species is a single short flash, but the time they allow to elapse before making it is the critical characteristic and a searching male will take no notice of a reply unless it comes at the correct time interval after he has concluded his call.

Once these codes have been deciphered, it is easy to replicate them with a small torch and persuade a hopeful male to fly in and land on your finger. And that trick is also played by the female fireflies. Unusually among fireflies world-wide, the females of some North American species are carnivorous and a substantial element in their diet is the abundant small males – of *other* species. In areas where several species occupy the same territory, the females will give their welcoming response not only to males of their own species but to different ones. Their answer is neither accidental nor haphazard, for the females recognise the species of other flashing males and modify their replies to match their code. Some can give the appropriate answer to at least five species of males and will switch from one code to another as different individuals fly by. And when a male is deceived and lands beside one of these large duplicitous females, he finds not a receptive mate but a predator. He is grabbed and eaten.

In New Zealand, another insect, a fungus gnat, also uses light to attract its prey. Its larvae live in many places – beneath bridges, under the overhangs of damp sheltered banks, within hollow trees – but they are most famous for gathering in millions within caves. Each of them secretes a tube of clear mucus which it suspends horizontally from the ceiling of the cave within which it lives. From this it hangs several dozen threads, each beaded with globules of glue. As it sits within its translucent home, it glows with a steady unblinking light so that the whole ceiling shines like the Milky Way. The light attracts midges and other night-flying insects which become entangled in the threads. The larvae above, when they sense the vibrations caused by the trapped insect, bite a hole through their tube and haul up the thread with their mandibles, like fishermen reeling in their catch.

Bioluminescence is comparatively rare on land, but in the sea it is widespread. If you dive in the clear Caribbean waters of the San Blas Islands off Panama at certain times of the year, green spots of light sparkle around swimmers, whether human beings or fish. The local people call the organisms responsible fire-fleas. They are minute crustaceans known to scientists as ostracods. They are about the same size as a flea with shrimp-like bodies enclosed within a delicate transparent carapace, not dissimilar to Daphnia, the familiar genus of water-flea that swarms in fresh-water ponds. Those displaying around you are using their light to protect themselves. The sudden spurt of luminosity in the dark water so startles a predatory fish that even if it has already snapped up the fire-flea, it may swiftly disgorge it. The fish may also find the lights flashing all around it very unwelcome, for they make it conspicuous to other bigger predators. It may therefore decide to swim swiftly away into the safety of darker waters and leave the fire-fleas unharmed.

But the fire-fleas also use their light in another way and one which closely parallels the fireflies. About an hour after sunset, as the last glimmers of twilight fade, the male fire-fleas begin to display. Small though they are, they can move swiftly through the water and as they go they emit a train of flashes. One species leaves behind it a long train of equally spaced spots of light. Another shoots up vertically, making flashes with accelerating frequency, each of which lasts several seconds. Yet another, like the South-east Asian fireflies, flashes synchronously. The males, more than a metre apart, move through the water turning on and off in precise unison. Nearly seventy species in the region have been studied and the complexity of their displays rivals that of terrestrial fireflies. Doubtless the purpose of many or all of these displays is to attract the females, but the details of their courtship have not yet been fully unravelled.

The fire-fleas create their luminescence with the same kind of chemical reaction that is so widely used on land. A luciferin is produced by a gland on one side of the mouth and a luciferase by one on the other. As both squirt into the water together, they react with oxygen to release light. Many marine animals, however, use an entirely different method. They employ cultures of luminous bacteria which they maintain in special organs within their bodies. This technique has its limitations. The flash of a firefly's luciferin is within its control. Bacteria are inclined to luminesce when it suits them rather than their host. The flashlight fish, a small creature about the size of a sardine, keeps these bacterial cultures in pouches on its cheeks which are covered by a patch of skin. This can be pulled up like an eyelid to conceal the continuously glowing bacteria when the fish decides it would be safer not to be noticed. The deep-sea angler fish keeps its cultures in a small bulb at the end of a long spine that droops in front of its mouth. The bacteria there glow only if they are kept supplied with oxygen-rich blood. The angler is able to contract the tiny blood vessels supplying them and so can douse its light when it wants to.

Around 90 per cent of all the organisms swimming in the black middle waters of the ocean utilise light in one way or another. Our ability to observe the behaviour of animals at these depths is still very limited and our knowledge of just how these living lights are deployed is often little more than guesswork. The flashlight fish seem to use them to maintain contact with one another as they swim in a shoal and then turn them off when a predator approaches. The angler fish uses its bulb as a lure, exploiting the strange attraction that all animals in the sea have for light. Some squid, when attacked, discharge a cloud of luminous mucus which hangs in the water and forms a veil behind which they can make their escape. And many organisms of all kinds, with their arbitrary species-specific codes of flashes, must certainly be using them to send messages to prospective mates.

Other marine organisms communicate with sound. A human diver hears little of it, for the air in the outer passage of our ears acts as a sound block, but if you lower an underwater microphone into the sea, you will discover that it is full of noise. Trigger fish grate their teeth together, herring fart to each other at night to keep the shoal in place, sea-horses rub their heads against the spines on their back and pistol shrimps deliberately dislocate their claws, making noises like gunshots. When a spiny lobster is threatened by a conger eel, it rubs its stony antennae along the toothed spike that projects from its head between its eyes, creating a rasping noise, and all the lobsters within fifty metres rush to take shelter in their holes.

San Francisco Bay is fringed by a number of small harbours where lines of houseboats float at permanent moorings. In the late 1970s, a programme was started to clean up pollution in the Bay. As it took effect, so the houseboats in the northern part of the Bay became very fashionable homes. But then their attractions were seriously marred. Night after night, those sleeping on board were kept awake by a loud and persistent hum. Some people maintained that it was caused by an electrical power-line that had recently been laid across the Bay and wanted to sue the electricity company. Others thought that the sewage works that had only recently been prohibited from discharging effluent into the Bay were secretly and illicitly pumping it in again under the cover of darkness.

It was a scientist from the San Francisco Aquarium who found the culprit – toadfish. The recent cleaning of the waters of the Bay had made it possible for them to colonise it. The males had established themselves among the rocks beneath the houseboats and started to sing by vibrating their air-bladders. Once they began, they continued for an hour or more. The hulls of the houseboats acted as sounding boards, amplifying their song into the drone that many people found so intolerable. A female toadfish, however, would find it irresistible. She would swim towards her singing male who suddenly grabbed her in his jaws, dragging her into a crevice and they would spawn. Since there was little the San Franciscans could do to dislodge the fish, they decided to celebrate them instead. Each year during the breeding season, the people of the Bay held a toadfish festival, until 1990 when declining toadfish numbers sadly led to the end of the celebrations.

The richest vocabularies of underwater sounds are those used by the members of the whale family. It has long been known that dolphins are very vocal animals. Those kept and studied in oceanaria have tamed so well that investigators have been able to work out the meaning and function of their calls in detail. Quite apart from the sounds they use in their sonar navigation systems, which have been brought to such perfection by the poorly-sighted endangered river dolphins, they also produce a great variety of whistles and grunts. These they use to indicate their moods and reactions in much the same way as a dog will bark with excitement, growl in anger and yelp with pain. And they do more still. While they are making such noises they can also simultaneously whistle. This sound is the animal's call-sign. Each is so characteristic that even the briefest snatch, half a second long, can be enough to enable one animal to recognise another.

Research in the clear shallow waters off the Bahamas with a free-swimming school of dolphins a hundred or so strong has extended our knowledge of dolphin communication still further. Not only does each dolphin have a vocabulary of about thirty different vocalisations but it can modify the significance of each by the posture it assumes while making it. A particular sound uttered while swimming will have a different meaning if the animal is also nodding its head at the same time. The signature whistle is not merely a statement of identity. It can also be used by other animals to attract its owner's attention, as though calling its name. And a young dolphin develops its own whistle which, while it is unique to itself, nonetheless bears a resemblance to that of its mother, just as a human child's looks may resemble one of its parents. Human-produced ocean noise – mainly from ships' engines – is hampering the ability of dolphins to communicate. In response to the sonic interference, the dolphins are changing how they call, altering the frequency. Sadly, this leads to the calls being less informative, limiting their social function and potentially affecting the social behaviour of these animals.

In addition to sounds which even human swimmers with their impeded ears can hear underwater, the dolphins also use their ultrasonic system which we cannot hear at all. Individuals can almost certainly sense an ultrasonic beam if it strikes them, and seem to exchange glances ultrasonically just as we do visually. They also communicate by touch – nudging, stroking and smacking one another.

With such an extensive vocabulary using the four different senses of sight, touch, hearing and ultrasonic sensitivity, it is tempting to say that dolphins have a language. But language is more than words. It must also have a system of putting words together to create additional meanings. That is to say, it has to have a structure and a grammar. Dolphins do not have such a language. The only animals that have developed such a medium are human beings.

The huge cousins of the dolphins, the great whales, do not produce the clicks and high-pitched sounds that dolphins use, but they do sing. The songs of the humpback have been studied more intensively than those of any other great whale. They are made up of vast roars and groans interspersed with sighs, chirps and squawks. Each song can last as long as ten minutes. Having finished it, the humpback will repeat it, chirp for chirp and roar for roar, and go on doing so for hours on end. The singers are all male and each of them within one great community will sing a similar song. Those around Hawaii do so with one dialect; those in the Atlantic with a different one. They are at their most vocal when they are lying in the warm waters where they breed, but they also sing as they cruise on their long migratory journeys. As they travel so their song changes. New phrases are added, others are lost. But all the whales in the community change their songs in the same way. By the time they return to breed in the following year, the alterations are substantial. The function of these calls is still unclear. Even though the songs of all individuals in one community are much the same, the voices of particular whales have their own idiosyncrasies and are identifiable, so maybe they serve partly as locating signals. They may also be claims to ownership of feeding grounds and, since they are sung only by males, have some sexual function as well.

The blue whale has a special place as a communicator. Like the humpback, it too sings, producing deep lugubrious moans, each note of which may last for half a minute. It is the largest of all living animals, growing to thirty metres in length. With such a vast body, containing a huge larynx and immense lungs, it is able to produce very low notes indeed and give them a titanic intensity. Blasts of 188 decibels have been reported. That is the loudest noise made by any animal. It is far beyond the threshold of pain in the human ear and is comparable to the shattering roar of the rocket that launches the space shuttle. Blue whales may not have a very extensive vocabulary but they certainly are able to project what it is they have to say. Their gargantuan groans echo across the ocean basins from one shore to the other and their signals may be heard by strangers over a thousand kilometres away.

ELEVEN

Courting

When strangers on the lookout for mates meet one another, the first question they must resolve is whether they belong to the same species. Unions between different species are only very rarely fertile. A horse and a donkey can produce a mule; a lion and a tiger, a tigon, but both these are matings, usually brought about by humans, between species that would seldom if ever encounter one another naturally. With very few exceptions, mating between different species in the wild would be unproductive. So if the strangers are not to waste their time and energy, their specific identities must be quickly discovered. That is the first function of courtship.

Emperor moths deduce the answer from their mates' perfume when the two are still a kilometre away; fireflies can tell from their flash-codes when they are many metres distant. But many animals have to get close to one another before they can make a certain identification.

The commonest lizards in the southern United States are small agile creatures which, because their scales are elongated into backward pointing spines, are known as thorny lizards. There are many different species of them, several of which occupy overlapping territories. Different and mutually infertile individuals are therefore quite likely to encounter one another.

The males on a warm day are very active, scuttling after insect prey, and they challenge one another by bobbing their heads up and down. The action is made more conspicuous because their throats are a vivid blue. But each species does this at its own characteristic speed. And the female lizard, like the female firefly, will respond sexually only to the one that is made by her own species.

Birds must also show their identity cards at the beginning of their courtship. Birdwatchers generally know what these are because they too use them to distinguish between closely related species. They call them 'field characters' and often mention them in giving species their popular names. The birds themselves, during their courtship, go out of their way to draw particular attention to them. So among the tits, all birds of roughly similar size and shape, the crested tit erects his crest when he displays; the bearded tit puffs out the beard-like black patches on either side of his head; the blue tit, who has a yellow chest, swells that; and the great tit does the same thing because, although his chest is also yellow, it has a black stripe running down the middle of it.

Terns, to the uninformed eye, are all very much the same – sea-birds with pointed wings, pale bodies and, during the breeding season, black caps. An ornithologist distinguishes them by the colour of their beaks. The little tern's is yellow, the Arctic tern's red, the common tern has a red beak with a black tip, the Sandwich's is black with a yellow tip, and the black tern's beak is pure black from one end to the other. All terns, during their court-ship, to the convenience of both their mates and their human watchers, brandish these clearly labelled beaks in as conspicuous a way as possible, pointing them up, pointing them down and waving them about.

Other birds use their voices to proclaim their species. These also help the skilled naturalist. From them he can rapidly identify the songbirds hidden in a hedgerow long before he is able to see them. Indeed, in some cases, it is much easier to identify a bird by its song than by its appearance. It takes a very educated eye to distin-guish between the three kinds of little greenish-brown warblers that visit England each summer. But anyone can tell the difference when they begin to sing. One produces a series of liquid musical notes that get louder and rise higher until they reach a climax and cascade in a downward scale. That is the willow warbler. The second starts with a stabbing *stip-stip-stip* call which increases in speed and ends in a penetrating trill. That is the wood warbler. And the third and most easily identified of them all has a simple two-note call that gives it its name – *chiff-chaff*. If such differences enable a human being to identify a bird species, they must most assuredly enable the birds themselves to do the same thing.

But how does an individual bird know what species it belongs to? To a great extent that is inborn knowledge embedded in its genes, but at least in some birds their early experiences as chicks have a considerable influence. The process of imprinting that causes ducklings, as soon as they hatch, to follow their parents, leaves a life-long mark on them. If an egg of a Canada goose is placed in the nest of a mallard, the little gosling that hatches from it will happily join the flock of ducklings, following the adults about, swimming and feeding. At that time, it does not look very dissimilar. Two years later, when it has acquired its adult plumage, it looks very dissimilar indeed. But it cannot itself be aware of its own appearance. Now when it seeks a mate, it searches not for another bird like itself but for one that looks like the mallards that reared it. So if the youngster is male, he will indefatigably court female mallards up and down the river, but to no avail, for they all find his head-dipping un-mallard-like displays quite incomprehensible.

A correct answer to this initial courtship question about identity is not necessarily enough in itself to lead to mating. A male, who in most species is the one who makes the initial enquiry, may have to persuade his prospective mate that he has more to offer than any of his rivals. Damselflies lay their eggs in water and are seldom found far from it. But in central America, the biggest of all damselflies, a giant with wings nineteen centimetres across, has managed to colonise the rainforest by using the tiny pools of rainwater that collect in the stumps of trees and the empty sockets left in a trunk by a fallen branch. There are not many of these and the males fight to take possession of them, diving and buzzing at each other until one is driven off. The victor then circles around in the air above his private pond, beating his large white wings with their prominent blue blotches at the tips, in a curious whirling motion that gives him the nick-name of 'helicopter damselfly', advertising the property he has at his disposal. A female ready to lay will not mate with any male she might meet in the forest. He has to be one who can provide her with a home for her eggs.

Female red-backed salamanders that live in North America choose their mates in another way. Each male, at the beginning of the breeding season, lives in his own burrow and each has the habit of leaving outside the entrance of his home a little parcel of droppings. Females looking around in search of a mate inspect these with great care. Sometimes they press their noses in them repeatedly, as though they were trying to discover what they contain. The salamanders feed on termites which are juicy and thin-skinned. That is their preferred diet. If they cannot find them, they make do with ants, which are not so rich and have hard indigestible skins. A female, by her probing of a dropping, can tell what a male has been eating, and she will prefer one who has been feeding on termites.

By making the choice, she ensures that her offspring have the strongest and most effective father that she can find. One cannot, of course, credit her with an awareness of such a purpose. She does so because her ancestors who first made such a selection, for whatever reason and however sporadically, produced more vigorous and effective young than those who did not. So by natural selection, the genes that cause individuals to behave in this way have become fixed in the salamander's hereditary make-up.

Similarly, many female birds in their courtship seem to be putting their prospective mates through a series of trials to ensure that the chicks have a father who not only gives them an effective genetic legacy but who also has the abilities to provide them with the food they need to build their bodies. So again, although one cannot say that the females have that conscious aim, their behaviour certainly has that result.

Terns live primarily on small fish which they catch by diving into the surface of the sea. A male, courting a female, brings her one or two, carrying them crossways in his bill. Male kingfishers behave in a similar way. The European bee-eater makes a particular song-and-dance about his presentations. Very soon after the birds arrive in the spring, having flown up from Africa, they begin to pair. As a prospective couple sit side by side, the male will take off, catch an insect that is flying past and then return to sit beside her, vibrating his fanned tail excitedly and calling loudly – in spite of the fact that his beak is full. Sometimes the female will take his gift with a show of indifference, but if she is about to accept him as a partner, she leans forward with her body in the horizontal position that invites copulation. The male immediately mounts and they mate even before she has had time to swallow her wedding present.

Female hanging-flies are particularly fussy about the size of such nuptial gifts. They are large insects with four narrow black-tipped wings, superficially similar to crane-flies or daddy-long-legs, and spend much of their time suspended by their legs beneath leaves. The last joint of their hind legs can be snapped forward like the blade of a jack-knife and with these they catch flies. A courting male, having captured one, will make a short flight, releasing a special scent, before returning with his prize to his position beneath a leaf. A female, attracted by his scent, settles opposite him. He then holds the fly towards her. She inserts her mouthparts into it and begins to feed. Without relaxing his hold on his gift, he then brings his abdomen forward and attempts to mate with her. If his offering is a small one, or if it is an insect such as a ladybird that she does not fancy, she will curve her abdomen away and after five minutes or so abandon him and his gift. If, however, the fly is large and to her liking, she will continue to feed for as long as twenty-five minutes while the male, having achieved a connection, slowly transfers his sperm into her body. Once copulation has finished, the male and female will tussle over possession of the prey – the male is generally successful in eloping with the partly-eaten corpse, which he can then use to seduce another female.

But all is not always quite so straightforward in the sex life of the hanging-fly. Catching flies is a risky business. Many hunters get trapped in spiders' webs. But some find an easier way to get the gifts they need in order to mate. One will land alongside another male that is proffering a fly in the hope of attracting a female. The newcomer lowers his wings, mimicking the gesture that a female uses to indicate her acceptance of copulation. Sometimes the hunter is not deceived and flies away, taking his present with him, but on two-thirds of the occasions, he offers his fly to the bogus suitor and attempts to mate. The mimic curves his abdomen away from that of the hopeful hunter, so delaying the discovery of his true sex. After another two minutes or so, the hunter gives up and tries to reclaim his fly. The two wrestle with it. Around one time in three, the mimic wins possession and flies away with his stolen goods.

There are other ways in which females may test the abilities of prospective males as providers of food. Eagles, if they are to be effective hunters, must be powerful and skilled flyers. An African fish-eagle courting a female on the wing demonstrates his aeronautical abilities in a dazzling way. He flies with her to considerable altitudes and then climbs still higher so that he is a little above and behind her. Then he dives down steeply towards her. As he approaches at speed, she makes a half-roll so that she is flying upside down and, just as he comes alongside, the two clasp talons. With feet interlocked, they cartwheel downwards, over and over, until still some distance above the ground, they disengage and flap away together.

In some species, courtship displays which may once have provided practical proof of a bird's virtues as a mate, seem to have become ritualised into purely symbolic acts. Great crested grebes, which start their complex dances with head-shaking displays, erecting the species-indicative crests and tufts on either side of the head, end by presenting to one another, not fish on which they feed, but fragments of weed. The billing and cooing of doves in which the pair fence amorously with their beaks can be interpreted as a ritualised presentation of food. But sometimes the characteristics that impress mates seem to have nothing whatsoever to do with their capacity to provide a home, catch food, drive off enemies or any of the other skills to be expected from a good mate and father.

The ability to sing a beautiful and complex song is of no great help in such matters. Nonetheless, that is the accomplishment by which many female birds distinguish between their suitors. The songs of males, when they are establishing their breeding territories, may undoubtedly serve as threats driving other males away. But they are also important elements in courtship. Sedge warblers, for example, sing lustily when they first take up residence in a territory and produce a particularly long song. But its complexity varies between different males. Careful observations and recordings have shown that those males with the most complex songs acquire females before those with simpler songs. Furthermore, as soon as the male has paired, he stops singing. If thereafter he needs to drive away an intruding male, he does not do it with song but with visual threats and physical violence.

The nightingale's abilities as a singer have become famous for he performs at night when most other birds are silent. It may be that he does so in order to call down females who travel up from their southern winter quarters under the cover of darkness and are passing overhead. He certainly deserves his fame for he produces a glorious sequence of deep throaty chuckles and high whistles, trills and flourishes which can last for many minutes on end. As the weeks pass he slowly increases the length and complexity of his arias, but after he has paired and the female has laid, he stops his serenades.

Visual splendour is hardly of value either, when it comes to raising a family. Yet female birds select their partners on this basis as well. Feathers provide the ideal means with which to create dramatic and spectacular effects. They are light, easily erected and easily folded away and, even in the normal course of events, are regularly moulted and renewed. Male birds exploit them to the full in order to appeal to the predilections of their females.

Many ducks, when the breeding season approaches, shed the drab feathers they wore through the winter and grow a very different set indeed. A small Chinese wood duck with a grey head and dappled brown chest puts on a particularly elaborate costume. The top of his head turns glossy green. A ruff of pointed feathers appears around his neck and, most extraordinary of all, triangular sails sprout from his wings. He has assumed all the finery appropriate to his name of mandarin.

Male mandarins display in groups in front of the females, dipping their beaks in the water, stretching and arching their necks and turning their heads to touch, with the tip of the bill, the inner surface of the erect sails on the wings, usually choosing the one on the side nearer the females they are trying to impress.

The mandarin has a close relative that lives in North America, the Carolina wood duck. During the winter its relationship is obvious for it looks very like a mandarin, but when the breeding season comes, the Carolina is also transformed. But in a very different way. His wings become a metallic blue-green and do not develop sails; his neck does not carry a brown ruff but turns a rich purple edged with a white line; and his eye which, like the mandarin's, had been dark becomes a brilliant red. The extreme difference between the breeding plumage of these close relatives emphasises how arbitrary the predilections of females may be.

Whether a female during courtship forges a link with a male that will last after copulation depends on whether two parents are necessary to bring up their young to independence. The female of most bird species, by herself, is unable to incubate her eggs and collect sufficient food for herself and, later, her chicks. These jobs require two adults. Ninety per cent of all birds therefore are monogamous, the pairs remaining together until their chicks are reared.

This is not the case in the animal kingdom as a whole. Many species give no parental care whatever to their offspring and even among those that do so, the females are well able to look after their young single-handed. For such species, therefore, there is no need for a long-lasting pair bond. Females and males make very different contributions to the creation of the next generation. A female mammal or bird can, at the most, produce only a few dozen eggs each season. A male, on the other hand, produces sufficient sperm to fertilise thousands. The number of young he fathers depends on the number of females with whom he mates.

Many males are therefore polygamous and fight other males in order to get as many mates as they can. As such species evolve, so these males become bigger and more powerful. Some such as bull elephants and Hercules beetles battle to claim any female that turns up. Others accumulate harems as do elephant seals and moose. For them courtship, in the sense of persuading their much smaller females to accept copulation, scarcely exists. Physical force gains them what they seek.

Polygamy also occurs among a small number of birds. These are exceptional in that the food that they eat is easily and quickly gathered and that their chicks need minimal care after hatching. Such females therefore are able to incubate the eggs and rear their chicks unaided. Male polygamous birds do not herd their females together with sheer force, as mammals do. Instead they attract them, one at a time, by taking to extremes the inducements practised so effectively by monogamous males — the quality of their songs and, particularly, the splendour of their adornments.

The male lyrebird of Australia has one of the most complex songs of any bird in the world. In addition to a large vocabulary of his own, he includes in his performances accurate imitations of many of the other birds that live in the woodlands around him, and even of human-produced noises such as the click of a camera shutter or, tragically, the sound of chainsaws cutting down the forest. One individual is credited with mimicking sixteen other species. And he sings so loudly that he can be heard, when conditions are favourable, over a kilometre away. As if that were not enough to impress a female, he has also developed very elaborate display feathers. Those in the centre of his tail are extremely long and the vanes on either side of the quills have been skeletonised to form ranks of separate silver wires. These plumes are framed on either side by a pair of broad feathers, marked with brown diapers and gracefully curved like the arms of a lyre. He gives his performances on courts over a metre and a half across which he builds by stamping down the vegetation where necessary, digging out roots and kicking up the earth. He may construct as many as twenty of these in his territory and during the breeding season he visits them in turn, singing loudly and slowly revolving with his tail plumes thrown forward over his back and head so that he is almost entirely concealed beneath a quivering veil of white filigree. The females, who lack his long tail, tour these courts listening to the various rivals, contemplating their dances and eventually selecting one with whom to mate.

As soon as copulation is over, the female goes off to lay her eggs in the nest that she has already made for herself and the male resumes his singing and dancing. During the season, he spends as much as half the hours of daylight giving these displays. The females come into breeding condition at different times over a period of seven weeks, so if he is a superior performer, he may succeed in mating with several of them.

The variety and magnificence of the costumes developed by birds for such competitive displays beggars the imagination. The tragopan pheasant not only has deep crimson plumage, spotted with large silver discs, but two fleshy ultramarine horns that he can erect on his head and a lappet on his breast which, at the climax of his display, he inflates into an electric blue balloon, patterned with scarlet.

The Victoria's riflebird of northern Queensland, in repose a handsome if shy bird with an iridescent triangular breast-shield, in display becomes transformed into an incarnation of ecstasy, his shield spread into a shimmering line of colour, his wings fanned and held vertically to frame his head, which he lowers and moves dramatically from side to side.

Lawes's parotia, a thrush-sized bird of paradise, black but for a green breast-shield, when he dances on his court in the New Guinea forest, extends his flank plumes into a kind of crinoline so that his feet are almost hidden and he becomes a black cone. He stretches up his neck, hops slowly from side to side and then, pulsating his neck-shield, he starts nodding his head from side to side so that the six long-quilled feathers that sprout from his forehead vibrate from side to side until they are lost in a blur.

The male blue bird of paradise has sapphire wings and white skin above and below his eyes so that he appears to be wearing old-fashioned flying goggles. During the breeding season, gauzy plumes, the same colour as his wings, sprout from his flanks and lower breast. His display court is not on the ground but on a particular branch of a tree. First he calls a few simple bugle-like notes. Then, gripping the branch firmly with his feet, he topples backwards and, hanging upside down, distends his breast plumes into an azure fan. A scarlet line that, in repose, stretched across the lower part of his breast, now expands into an oval disc which pulsates in size. Two long tail quills, naked of vanes, each with a blue spatulate tip and twice the length of his body, wave above him. When the female comes close, his display increases in intensity. Still upside down, he tilts himself sideways towards her, thrashes his tail quills feverishly from side to side and produces as strange a noise as comes from any bird's throat, an extraordinary mechanical whirr like an electric hand-drill, throbbing with the same beat as his pulsating plumes.

Effective though extravagant feathers like these may be in attracting females, from other points of view they are a considerable impediment. They trail behind a bird, making him less agile in the air and more conspicuous to his enemies. So at the end of the breeding season, a male bird who developed them for his courtship now moults them, just as the bull moose drops his antlers. He then grows them afresh the following year. This clearly consumes a great deal of bodily energy.

One group of birds, the bowerbirds, which are closely related to the birds of paradise, have evolved an equally spectacular courtship technique which avoids all these disadvantages. They live in New Guinea and northern Australia. Although some have crests, they have no large elaborate plumes. Instead, they make their displays with brightly-coloured objects which they collect from the surrounding forest. Each bowerbird species has its own particular aesthetic taste and each builds its own individual design of show-case in which to display its treasures.

There are four main types of these constructions. The simplest is made by the toothed bowerbird, also known as the stage-maker, a plain brown creature about the size of a jackdaw. He clears an area of the forest floor up to two and a half metres across, meticulously removing all debris so that it looks as though it has been swept with a brush. He even cleans the bases of any saplings that may be standing in it. Then he cuts leaves from particular kinds of trees, sawing through each stem with his toothed bill. It may take him as much as a quarter of an hour to collect just one. These he uses to carpet his court, laying down each one with its underside uppermost so that it shines palely in the gloom of the forest. By the time he has finished, he may have arranged as many as a hundred leaves on his court. Every day he throws away those that are withered, dropping them in the forest just beyond the edge of his stage; every morning he brings fresh ones to replace them. Once his stage is furnished to his satisfaction, he sits in a tree above it, singing loudly. If a female appears, he flutters down to the stage and displays to her by crawling slowly across it, crouched low, flicking open his wings and jerking his tail.

Archbold's bowerbird constructs bowers of another kind. He clears avenues across the forest floor which he decorates with piles of beetle wings, snail shells, berries and chips of amber-coloured resin from tree ferns. He even, significantly, will glorify them with the moulted plumes of birds of paradise.

A third group builds bowers with two parallel walls of twigs over thirty centimetres high and as far apart, interlacing the twigs that form them with as much care as any bird building a nest. At each end of this corridor, he piles his jewels. Spotted bowerbirds favour white things – pebbles, shells, weathered bones, small crystals. Satin bowerbirds prefer blue ones – parakeet feathers, berries and, where they are near to human settlements, blue plastic. The satin bird goes to even further lengths to beautify his bower. With his beak he paints its inner walls with pulped blue berries.

The fourth type of bower is more elaborate still. Sticks are built into tall constructions around the trunks of saplings. MacGregor's bird erects a single tower, with a circular low-walled track in the earth around its base, round which he dances as though circling a maypole. The golden bowerbird builds two towers, using a pair of neighbouring saplings and building up a rampart of sticks between them to form a saddle which he decorates with tufts of pale lichen. And most complex of all, the gardener bowerbird uses the trunk of a sapling as a central pillar to support the roof of a conical hut, a metre high and a metre and a half across. In front of its entrance he lays a lawn of plucked green moss, and on this he carefully arranges piles of flowers, fruit and brightly coloured fungi.

The theory that these bowers serve the same purpose as courtship plumes is supported by the fact that there is a close and inverse correlation between the two. This last group of species are all closely related to one another. Those that build the simplest bowers, the maypoles, have large yellow crests; the golden bowerbird with its double tower has only a small one; and the gardener, which makes the most elaborate tower, has no crest at all.

The bowerbirds, the lyrebirds and the pheasants all establish their display courts some distance away from those of their rivals. But other polygamous males find it expedient to bring their courts together in one large arena where they can display competitively alongside one another.

Several close relatives of the blue bird of paradise do this in New Guinea. The lesser bird, which has sulphur yellow plumes, gathers in a tree in groups of up to ten, each with his own stretch of branch on which to call and strut with his plumes erected and trembling over his back.

On the American prairies in spring, sage grouse gather in groups of several dozen, fanning their tails into spiked circlets like sunbursts and inflating air-sacs in their throats which they suddenly contract so that the air within them is expelled with a crack that can be heard half a kilometre away.

In the South American rainforests, cocks-of-the-rock, brilliant orange, each with a permanently erect semicircular crest tipped jauntily like a cocked hat over his forehead and beak, sit perched low in the trees in assemblies of forty or fifty. Sometimes they form groups squawking and squabbling among themselves. Sometimes they are more scattered, each sitting above the small patch of cleared ground that is his own personal court. But as soon as the brown female appears, they all flop down to their own particular places and crouch there, wings splayed and their heads held on one side so that their orange crest is horizontal.

The calf bird is one of the very few species indulging in these arena displays that does not have bright feathers. Although it is a close relative of the flamboyant cock-of-the-rock, it is a dull brown and both male and female are similar in appearance. Half a dozen of the males gather together in a tree. After a period of solemn silence, one will take a deep breath and, raising himself up on his legs, emit a long lugubrious *moo*, rather like the call of a calf. As soon as he finishes, a rival does the same thing. Each bird carefully takes his cue from another so that the calls do not overlap each other. And instead of indulging in passionate trembling struts, the peak of a calf bird's display is to fix his rival with an unblinking stare and very very slowly lean forward until his body, his outstretched neck and head are all horizontal. He remains frozen in this position, sometimes for ten minutes, on occasion for as long as three-quarters of an hour.

Many of the birds competing in communal arenas display whether or not there is a female in the neighbourhood. It is as though they shared the view of those human sportsmen who maintain that the important thing is not to win but to take part. That certainly is the impression given by the astonishing performances of the manakins.

These South American birds are a little smaller than sparrows. Although some are brightly coloured, they compete primarily in the complexity of their dances. The black and white manakin clears a small area of all litter and then performs frantic gymnastics in the branches of the bushes and saplings that stand in it, leaping to and fro between two upright stems, swaying his body with whirring wings, scuttling down a branch head first with such swift foot movements that he seems to be sliding. All this is done to the sound of high-pitched calls and cracks, snaps and grunts, made mechanically by special quills.

The blue-backed manakin is particularly strange in that, in order to raise the choreographic complexity of their displays, two males collaborate like circus acrobats. The dominant bird starts by summoning his assistant with a call. When this junior bird arrives and sits beside him, the two issue an invitation to females with a rather longer call. This is in fact a duet, initiated by the senior bird and joined by the junior within a fraction of a second. When a female appears the two go down together to a special perch close to the ground and begin to bounce up and down alternately, rising only a few centimetres and accompanying each jump with a call. If the female flies down to them, then their acrobatics increase in intensity. Facing the female, one behind the other, with their bodies parallel to the horizontal branch, the front bird leaps into the air, hovers momentarily on beating wings and then flies slightly backwards so that he lands behind the second bird, who then edges forward in a behaviour resembling a dancer's 'moonwalk' and performs the same movement himself. As excitement mounts, the two do this with increasing speed, the landing of one being the cue for the other to take off, until the two birds are whirling like a Catherine wheel. If the female is still sitting in front of them, presumably transfixed by this extraordinary display of virtuosity, the dominant male calls two sharp notes and the junior bird, having played his part, leaves the court. The dominant male, now left alone with his female, starts to court her directly, fluttering round her and repeatedly alighting beside her, vibrating his slightly opened wings and lowering his head so that his scarlet cap is presented squarely to her with his brilliant blue back showing above it. If after all this performance she is still there, then at last he mates with her. Nor is this the most complex of manakin dances. The blue manakin of south-eastern Brazil gives a similar performance, but with teams of males that may include three or even more birds sitting in a line on a display perch and taking it in turns to bounce into the air.

Such communal displays were first described from somewhat less glamorous birds, the ruffs, small Scandinavian sandpipers the males of which develop large and variously coloured feather collars and display in groups on marshes and flooded meadows. Their assemblies were given the Swedish name of 'lek', meaning a playground, and that term is now used for all such arena displays.

It is not only birds that form leks. Even insects do so. Orchid bees, brilliantly coloured with a metallic sheen to their bodies, summon their females, not with sound as birds do, but with the usual insect method of smell. The males chew orchid flowers, stow the product in pouches on their legs and fly off to their arena, usually a sunlit patch on the trunk of a fallen tree. Each one marks his court on it with this perfume. Then they dance, touching the bark with their heads, raising their tails and buzzing their wings to make a noise that can be heard four metres away, and finally rising three or four centimetres into the air to hover momentarily before settling down again. The female bees visit these communities and select their mates from among the members.

Mammals too have their leks. The hammerhead bat, which is the largest of all African bats, living in the wetter western part of the continent, has a fair claim to have the most bizarre of all mammal faces. The great heads of the males have grossly enlarged mouths and muzzles with puckered lips and huge cheek pouches. These features, in the past, have been regarded as special adaptations that enable the bat to take a whole fruit into his mouth and suck it without spilling any of the juice. The problem with this explanation is that the females, which are only half as big, have heads without any of these grotesqueries, very like the foxy faces of other fruit bats. Yet there is no known difference in diet between the two sexes. The real reason is that the males compete with one another for the attention of females in leks.

The male bats, twice a year at the beginning of the dry season, assemble in the forest, usually along the bank of a waterway, hanging beneath the trees in a long spaced-out column, two or three individuals wide and fifteen metres apart from one another, for distances of almost two kilometres. When they first arrive, they squabble for positions, but once that is settled, they take up their accepted places every night and start a loud metallic honking, uttering several notes a second and flapping their wings twice as fast. The smaller females fly along the column, hovering in front of different males. When a female appears, the male responds by clasping his wings tightly around him and increasing his frequency of honks until they become almost a buzz. She may require several inspections of any particular male before she makes up her mind, but when she finally decides, she hangs up alongside the male of her choice, copulates and within thirty seconds leaves.

The loudness of the males' calls seem to be a critical factor in her decision. Certainly the need to out-honk a rival has led to a huge increase in the size of the male's larynx which fills most of his chest and the development of extensive amplifying air passages in the nose. So the hammerhead's astonishing physiognomy is comparable to the enlarged plumes of the bird of paradise – even if it is hardly so beautiful.

A close relative of the hammerhead, the Wahlberg's epauletted fruit bat, also indulges in competitive display. The male's calls are not so loud as the hammerhead's. Instead, they attract the somewhat smaller females with visual displays. The 'epaulettes', which give them their name, are tufts of long white hair that sprout from a pocket on each shoulder. As a female approaches, a male does his best to entice her by everting these pouches so that the hairs are erected into spectacular white globes.

Topi are large African antelope with short gracefully curved horns and a particularly glossy chestnut coat marked on the upper legs with purplish black. The way they behave during the breeding season varies. Some males, living in country where there is good grazing and scattered bush, set up their own individual territories which they guard against other males and to which females may be attracted because of the quality of the feed there. But others, living on the open plains, assemble to form leks.

Up to a hundred males travel to an arena, which is in a place that has been used for this purpose year after year. There they fight among themselves, laying claim to particular patches of ground, their courts. The most powerful males take the centre courts in the middle of the display area. Individuals are now about thirty metres apart, each with his own patch of territory, trampled bare and marked with piles of his own dung. Often a court is centred around an old broken-down termite hill, on which the buck stands for much of the time, his forelegs on top of the hill so that his shoulders are high, with his head raised like a sentry in a ritualised pose.

Occasionally, he leaves it and gallops around his little territory. When neighbours meet, they face one another across their mutual frontier, throwing up their heads and then lowering them as though they were about to charge. On occasion, they drop on to their knees and remain there locked in motionless threat.

As the females become sexually receptive, they visit the lek. Each makes her way gingerly between the junior males, towards the centre court. The juniors try to mate with her as she passes but she easily avoids them and they do not leave their individual stations. So at last she reaches the most senior male in the centre and there she mates.

The central courts are undoubtedly the best positions for they are the safest. Standing out on the open plain, the topi are very vulnerable. Lions and hyaena, hidden by the thick grass that fringes the whole arena, can creep up and pounce on a male standing out in the blazing sun, tired out by his cavorting and not at his most alert. Males on the outer courts, therefore, are at considerable risk. Not only that, but they seldom if ever get a chance to mate with the females. It seems that they are the losers from every point of view. The question must therefore be asked, why should any male – topi or orchid bee, sage grouse or manakin – take part in such gatherings? Would it not be better for him to set up by himself?

In the case of the topi, the junior males are slightly smaller than both the senior male in the centre of the lek and those that have set up independent territories in richer country. For them the lek may not offer a good chance of mating but it may be their only one. There is the possibility that they might catch a female on her way in. It could be that several females arrive at the same time and the senior male will be unable to pay attention to them all. In cock-of-the-rock and sage grouse leks, birds on the outer courts may have a reasonable chance of graduating to the better positions. These are held by older birds and they, because of their additional years and the effects of their exertions, may have only a few seasons ahead of them. So there may soon be vacancies which junior males may claim. Similarly, a young blue manakin may join a display team for the chance of inheriting the all-important perch if and when the senior partner falters.

The attraction arenas have for females is easier to understand. The commotion and spectacle created by so many dandies must be irresistible. They can hardly be expected to pay much attention to an individual male who might be displaying by himself. All the excitement is at the lek. There they have the best choice and will be able to select the finest male around. And why should they not have the best, since they are the ones who will probably do all the work, raising the family when the party is over.

TWELVE

Continuing the Line

The process of bringing egg and sperm together and creating a new generation can be fraught with great difficulties. It may have to be timed with precision; it may render those involved temporarily defenceless; it can risk serious injury; and in some cases it brings certain death. It is the final trial of life and its ultimate triumph.

The palolo reduces its mating risks to a minimum by exploiting an ability we still find inexplicable. It is a worm thirty centimetres long that lives in billions in the reefs of Fiji and Samoa in the western Pacific, burrowing with its strong mandibles through the stony skeletons of corals and eating the small polyps. Inside its tunnel, it is safe from all predators and it seldom emerges.

Its body is divided into segments, like an earthworm's, and each contains a set of the organs necessary for life. But sex glands develop only in those of its rear half. When breeding time comes, the worm projects its rear end out of its tunnel and breaks it off. This then wriggles to the surface and there releases its sex cells.

So the adult worm, still in its burrow, has succeeded in spawning without putting itself at risk in any way.

But the success of this technique depends upon its timing. If the worms are to achieve cross-fertilisation, then all of them must detach their rear ends simultaneously. And they do – at dawn on the first three days of the moon's third quarter in October, and then once again at the same time in November.

Palolo is greatly relished by both Samoans and Fijians and both people are able to predict the date when the worms will appear. The night before the rising is due, people from all over the islands travel down to the beaches. An hour or so before dawn, a few of the most eager will be wading in the darkness, searching with torches for signs. Even before the night pales into dawn, green wriggling strings materialise in the black water, spiralling upwards towards the lights. The call goes up that the worms have been seen and people who have been sleeping on the beach wade out, armed with nets and scoops. As dawn silvers the sea, the rising worms rapidly increase in numbers until great expanses of the water are covered in them. In a good year, they may form curds many centimetres deep. With shrieks of excitement and jubila-tion, the people ladle them into buckets. Big fish swim in, darting among the legs of the waders, frenziedly claiming their share of the bonanza. The thin body-walls of the palolo rupture in the waves and the eggs and sperm turn the water a milky blue-green. On the eastern horizon, the sun climbs out of the sea and within half an hour of the worms' first appearance, all is over.

Exactly how vast numbers of these organisms achieve their synchronisation is still debated. It cannot be that each worm has within it an internal clock which triggers action every 365 days because the moon's movements are not neatly synchronised with those of the earth, so the moon's third quarter in October arrives ten or eleven days earlier each year, until it slips back a month. Nor can it be that the worms judge the phase of the moon by its light for they spawn whether the sky is clear or completely overcast. One group of vigilant worms cannot be cueing others, for palolo on the reefs of Samoa and a thousand kilometres away around Fiji spawn at the same time. Furthermore the timing seems to be quite arbitrary, without any celestial or oceanic logic, for the Pacific palolo has a close relative on the other side of the world around Bermuda and the West Indies, and although it too spawns at the third quarter of the moon, it chooses to do so not in October but in July. The answer may not lie directly in the worms at all. Palolo spawning events are always preceded by spawning in corals, with the surface of the sea becoming awash with eggs, sperm and palolo segments. The corals, it appears, are able to track the growing period of darkness between sunset and moonrise that occurs towards the end of the lunar cycle. Coupled with other environmental signals relating to the time of year, the corals can identify when to spawn, and the palolo seem simply to follow suit.

Palolo worms are either male or female, but many sea animals are both simultaneously. Sea slugs, molluscs that have lost their shells and have developed a great range of beautiful colours on their naked bodies, are among them. They nonetheless go to some lengths to ensure that each individual finds a partner with whom to exchange sex-cells. One, Navanax, develops its eggs and its sperm at the same time. When two mature individuals meet, they obligingly alternate roles. First one behaves as a male, everting a long tentacle-like penis from a pore in its head and pursuing the other by following its trail of mucus. When it eventually catches up, the pursued lifts its hind end, allowing the pursuer to insert its penis in the genital pouch in its rear. The two then move along in tandem. After about ten minutes, they separate and change roles. Now the pursuer is pursued and he becomes she. Once again they copulate. They may keep up this performance for a long time, regularly changing roles. On occasion, they even form processions, the first in line playing female, the last acting as male, and those in between being both male and female simultaneously.

Land slugs are also bisexual. Some species start off as one and turn into the other as they develop. The dusky slug needs a sheltered moist site for its eggs and if necessary will fight a rival in order to get it. The two do battle by rasping at one another's sides. Size is very important in this duel and a small individual seldom wins. Delivering sperm, on the other hand, can be very effectively done even when small. So the dusky slug becomes sexually active as a male very early in its career and only when full-grown does it start to produce eggs.

More surprisingly perhaps, some fish can also change their sex. The little clown fish that gambol among the tentacles of anemones form small communities dominated by a monogamous pair. Young non-breeding individuals live in anemones close by the breeding pair, waiting their turn. If the dominant male dies, one of these will become sexually active and take his place. But if the breeding female dies, it is her partner, the bereaved male, who changes sex and takes her role, mating with a new male drawn from the ranks of the sub-adults.

Some species of wrasse which also live on reefs do things the other way round. They start as females and visit larger males who hold and actively defend territories among the corals. But as the small females grow, so they become big enough to set up a territory on their own account. They change sex, start fighting and, having established themselves, wait for a young individual, still operating as a female, to come and join them.

Many small animals – aphids and mites, wasps and termites – are able for many generations to dispense with the problem of copulation altogether. They can produce eggs which develop without any contact with sperm. The resultant offspring are clones, genetically identical with their single parent. Such an ability is of particular value when an animal needs to generate a large work-force in order to take swift advantage of a fleeting opportunity. So a single aphid can cause a plant shoot to be infested by a thousand carbon copies of itself within a few hours; a mite can duplicate itself to form a solid carpet covering all the available space on an insect host; and queen termites and bees surround themselves with whole armies of workers.

Many lizards reproduce themselves in this way as well. It is more difficult to understand why. At least twenty-seven species belonging to seven different families exist mostly, if not entirely, as females. None of these individuals, however, can reproduce herself in solitude. Each needs the encouragement and stimulation of another female, who plays the part of a courting male. Such a pseudo-male will later, with the help of another companion, produce her own clutch.

If this is possible, why should an animal involve itself in the difficult and dangerous practice of sexual congress? The usual answer given by biologists is that this process allows a reshuffling of the genes and, in consequence, produces variety in the offspring which enables evolution by natural selection to proceed. That is of value to a lineage of animals for it ensures that if the habitat varies, bringing slightly different opportunities and hazards, there will be individuals who are able to take advantage of the new circumstances. If all were identical clones, a single environmental fluctuation might kill them all.

Some scientists find this answer unsatisfactory. In stable circumstances, they say, the benefits of sexual reproduction are either non-existent or very small compared to the difficulties and dangers that animals must endure to find and copulate with a mate. How much more economic and efficient it would be for an animal to abandon its search for a partner and devote all its time and energy to producing young unaided. And indeed, this seems to have happened in the case of the microscopic bdelloid rotifer, a tiny freshwater animal. This group has reproduced asexually for tens of millions of years – although some species appear to occasionally produce males, most seem genetically incapable of doing so. Whatever the case, this arrangement has not hampered the rotifer, which has branched out into around 500 species and is found all over the world.

One explanation of why the rotifer's solution has not been more widely adopted is that there are very few circumstances indeed that are truly stable. Everywhere there are microscopic disease-causing organisms. These, because they reproduce at very great speed, are continuously evolving and diversifying and so are able to exploit weaknesses and vulnerabilities in their potential hosts. The animals they infect must therefore themselves be changing their genetic make-up if they are not all to succumb. The debate is still active, including the mysterious case of the rotifer. It is, however, the case that even those animals that dispense with sexual behaviour for many generations, revert to it on occasion, and that for the vast majority of higher animals, sexuality with all its hazards and complications is the only route to reproduction.

And hazardous it can most certainly be. It is particularly risky for animals who are hunters, for they are armed with teeth and claws and other weapons that, in a moment of careless passion, could kill. Spiders are one such group.

Copulation for spiders is a complicated business. The male does not have any special anatomical apparatus for introducing his sperm directly into the female. Instead, he spins a small silken napkin. On this he deposits a drop of sperm from the genital pore on the underside of his abdomen and sucks it up with his palps, feeler-like organs on either side of his head. He must then thrust one of these into the female's genital pore and squirt out the sperm, like liquid being expelled from a pipette.

His main problem, however, is not how to transfer the sperm but how to get close enough to do so without losing his life. His mate, after all, is armed with murderous poison-laden fangs; how can he let her know that he wishes to be a mate and not a meal?

Wolf spiders hunt by vision and have eight very good eyes that enable them to do so. A male wolf spider therefore uses visual signals to declare his identity and his intentions. His mate is marginally bigger than he is and he must be cautious. As he advances towards her, he rises high on his legs and signals with his palps which are conspicuously patterned in black and white, waving them jerkily up, down and sideways in a feverish semaphore, as though his life depended on it – which indeed it does. These are not only visual signs; as he moves his body, vibrations pass through the substrate to the female, producing a purring sound. He repeats these signals over and over again. If the female is not inclined to mate, she runs at him, just as she runs at her prey, and he retreats very rapidly indeed. But he is extremely persistent. If ultimately she relents, she conveys her assent by vibrating her front legs. Now he can venture forward with confidence. He clambers over her body with his head facing her rear, reaches round her abdomen to her genital pore and injects a drop of sperm. Then he may lean over the other side and give her a second instalment.

Pisaura, one of the European wolf spiders, does not embark on these delicate negotiations empty-handed. He first catches a fly which he gift-wraps in silk. As he gets within striking distance of a female, he rears up and leans backwards teasingly with his forelegs held vertically above his head, gripping his offering in his jaws. If she is to claim it, she has to reach up and collect it. As soon as she does so, he swivels round, ducks under her chest and while she is occupied unpacking her present, he delivers his sperm.

Other spiders who hunt in different ways have to conduct their courtship in a different manner. The huge hairy American spiders known as tarantulas do not have very acute eyesight. A male encountering a female starts by drumming on her body with his front legs. She reacts as she always does when alarmed, by lifting her forelegs threateningly. The male, however, continues to tap and stroke her soothingly. In response, she lifts her body still higher on her hind legs and opens her formidable curved fangs. A single downward stab from them could kill him. But he has special safety equipment. A pair of hooks on his front legs neatly engage with the fangs and hold them out of action. In this position, he brings his body close to hers, beats a final tattoo on her chest with his palps and then leans forward and achieves his purpose.

The European crab-spider is also short-sighted. His problem is particularly acute because his mate is very much larger than he is. He approaches her with extreme care, creeping low and moving slowly. When he is sufficiently close, he reaches forward and strokes her with his forelegs. As he continues with his caresses, he climbs up on her huge back, trailing behind him a line of silk. Back and forth he climbs over her mountainous body, securing his silken ropes on either side until he has tied her down to the leaf on which she sits. Only then does he gently lift up her abdomen and crawl beneath to mate.

Orb-weaver spiders are particularly sensitive to vibrations. When a female detects the struggles of an insect caught in her silken trap, she immediately rushes across and sinks her fangs in it. Many females spend all their time sitting on the web so the male strums on the outer strands of the web, using a special regular rhythm, clearly different from the irregular shakes caused by a struggling insect. He then clambers gingerly towards her. But he trails a safety rope behind him so that if she does not recognise him and attacks, he can swing swiftly out of danger.

The disparity in size between male and female is greatest in the Nephila spider of the American tropics. The male may have to deal with a mate as big as a human hand and a thousand times bigger than he is. But the disparity is so great that he is in no real danger, for he is below the size she considers worthy of attention. If she were a fisherman and he a fish, she would throw him back. When he finds her, he crawls over her immense body as she sits on her web and delivers his sperm, usually, it seems, without her even noticing.

Among mammals, it is usually the male who is the bigger of the two sexes. A bull elephant seal grows to three metres in length and can weigh two and a half tonnes. Females are only about half as long and a third as heavy. Mating takes place on the beaches where the pups are born. There are relatively few of these places and many seals who want to use them, so a male, if he is big and strong enough, can dominate a long stretch of beach and herd all the females on it into a harem. Then he fights with any other male who approaches. They rear up to one another, inflating the bladders on their noses, roaring ferociously and striking at one another with their long canine teeth until the necks of each are gouged and scarlet with blood. It is this premium on strength that has led to the males becoming so much bigger than the females.

A single dominant bull, a beach-master, may hold a harem a hundred strong. When a cow ceases suckling her pup, hormonal changes in her body stop the flow of her milk and then cause an egg to be released from her ovaries. Now she is sexually receptive. At this moment, she feels the urge to return to the sea. She has had nothing whatever to eat for the past three weeks while she was feeding her pup and now she is extremely hungry. Slowly she edges towards the water. The beach-master is quick to detect such a purposeful movement and hummocks thunderously across to her. Satellite bulls lying around the periphery of the harem at a respectful distance are equally aware of her movements and keep a close eye on her. Maybe more than one cow is on the move. If there is, then the beach-master will be in a quandary. Eventually he has to commit himself to one and lollops over to her. No matter how she tries, she cannot outpace him. He seizes her in his jaws by the scruff of her neck, almost crushing her with his weight. While he is so engaged, a junior male may see a chance and grab one of the other receptive females. Even if a female were to escape all the males on the beach and reach the sea unnoticed, she would still be pounced on by other bulls patrolling in the breakers.

The process of coupling seems so brutal and violent that anyone watching these dramas might conclude that the cows, running the gauntlet of the bulls, were doing their best to remain unmated. Whether that is the case in this instance or not, the females of most species are just as anxious to reproduce as the males, and many indeed go to a lot of trouble to advertise their availability.

Siberian dwarf hamsters are smaller cousins of the golden hamster from Syria that is such a popular domestic pet, and live on the bleak open steppes of central Asia. The breeding timetable of the female is necessarily very compressed, for her life is short and she has only a single brief Siberian summer in which to reproduce. Like elephant seals and other mammals, she cannot release an egg while she is producing milk for her offspring. She nonetheless manages to reduce the delay this could cause to her reproductive schedule by a carefully-timed publicity campaign.

The night before she gives birth to a litter, she marks the vegetation around the entrance to her burrow with strongly smelling vaginal secretions. Then she retires below ground to her nursery chamber. Meanwhile her scent wafts across the steppes. Males more than half a kilometre away are able to detect it. The following evening she gives birth. She then has three hours before she needs to produce milk for her young. In that short period she must mate again. By now males, forewarned of her condition, have arrived at the entrance to her burrow. Rapidly, she mates with one of them. Then she returns below ground to care for her new-born young. During the next eighteen days while she suckles them, a new brood will be developing within her womb. As soon as they are born, and before she suckles them, she will mate once again. By such carefully-timed programming, she may produce four litters in her short life.

A female elephant also broadcasts the news of her sexual availability. She does this not only with scent but also with sounds and in a way that ensures not merely that she gets a mate, but that she secures the biggest and the most powerful one available. She comes into season only about once in four years and then for only six days. The nucleus of elephant society is a group of mature females, usually sisters and daughters with their immature young. Wandering bulls pay regular visits to this group, smelling the cows to see whether or not they are in season. If one is, there may be a chase, but usually the bull copulates with a cow with little courtship or fuss.

But that is only the beginning of things. Once copulation is finished, the bull stands close by the cow as if guarding her. She now emits a deep rumbling noise. Some of its component wavelengths are audible to humans but some are far below the range of our ears and they travel for long distances across the African plains. A bull as much as eight kilometres away can detect them. Bull elephants are only sexually active for a short period of the year. Then a gland on either side of the forehead, the musth gland, begins to pour out a sticky secretion which forms an odorous dark stain down the side of the head. At this time, bulls are very aggressive and on the lookout for females. If the deep infra-sounds emitted by a female after copulation reach one of these, he will quickly set off to track her down. He may well discover that the bull who is guarding her is bigger than he is, and that is likely to be the end of the matter. But if he is the bigger one, he will drive off the smaller bull and mate with her himself. As soon as he has done so, the female once again repeats her deep rumbling call. This may be heard by another male who may prove to be even bigger. Many copulations later, as she approaches the end of her six days of sexual receptivity, she will have standing beside her the biggest bull for kilometres around. Only now does an egg within her move down from her ovary into her oviduct. Only now is she able to conceive.

Female lions, like elephants, live in family groups of sisters, daughters and their young. Males, either singly or in groups of two or three, settle in with them. But other males may appear and challenge the residents for the privilege of staying with the pride. Should the newcomers win the battle and take over, their accession is followed by carnage. The victorious males systematically kill the young suckling cubs. The lactating females with no hungry mouths pulling at their teats, stop producing milk and rapidly come into season again. The new males then mate with them.

The explanation of this behaviour, so horrifying to us, is that the male lion, like all individual animals, is concerned not with the good of the species as a whole but with the propagation of his own particular lineage, his own genes. Cubs fathered by others have no claims on his affections or support. It is only his own that he wishes to perpetuate. Since he is driven to behave in this way by the influence of his genes, it could be said that it is the genes themselves that are working selfishly to ensure their own survival.

Such killings by males of unrelated young occur among many animals. Langur monkeys behave in this way. Their social organisation, like that of lions, is based on groups of females with their young who are tended by a small group of males. Few of these manage to stay with the females for long. After two or three years they will be driven out by another male group and, once again, killings follow the take-over. Suckling babies are snatched when their mothers are momentarily inattentive and killed with a swift bite. You might think that a bereaved mother would have little to do with the murderer of her babe, but it is not so. Within days, sometimes within hours, she will copulate with him.

The need or opportunity for a male to destroy his step-children can only occur among animals who have a long gestation or whose young remain dependent for a considerable time. For most animals, this is not the case. Then the male's most important reproductive task is limited to ensuring that it is his sperm and no other's which fertilises a female's eggs. But even that is not easy to achieve. The male banded sunfish courts his female among the corals of a tropical reef with a great show of quivering fins and flashing colours. If he sees another male approaching, he darts at him aggressively, drives him off and then returns to curvette around the female, inducing her to spawn. But as the two come together, a third male suddenly appears from among the coral where he has been lurking unobtrusively, slips alongside the female, ejects his sperm over her eggs and darts away. Often the first male is so lost in his sexual excitement that he is unaware of what has happened. Over sixty species of fish are known to behave in this way.

When tricks like this can be played, it is not surprising that many animals go to great lengths to seize a female at the very first moment that she becomes sexually available. A female crab can only mate in the short period between shedding her old cramped shell and the hardening of her new larger one. A male, detecting from the release of chemicals from her body as she prepares for the process that this brief period of availability is about to arrive, will climb quickly on her back and cling there, fighting off all rivals, until that important moment.

Male Heliconius butterflies are equally attentive. They are not only able to recognise that a pupa hanging in a tree like a large seed contains an individual of their own species, but they can detect what sex it is. An un-emerged male is of no interest to them, but if there is a female within the pupa, they will cluster on it and settle in the twigs all around. Inside, the insect is in the last stages of her transformation from caterpillar to butterfly, her body hunched, her legs pressed tightly against her thorax, her wings crumpled and un-inflated. As the moment for her emergence gets nearer she begins to shudder. The end of the pupal case splits and she slowly starts to haul herself out. The males are now fluttering their wings with excitement. In some Heliconius species, they are so eager to copulate that they use the claspers on the tip of their abdomen to tear a small hole in the wall of the pupa through which they insert several segments of their abdomen tip. In this way they are able, as the end of the female's abdomen passes theirs, to mate before she emerges. In other species, they wait until her body is clear and then copulate with her even while her body is still drying and her wings expanding.

This one swift mating provides the female Heliconius with all the sperm she needs. She keeps it alive within her and draws on it to fertilise all the eggs that she will lay, a few a day, during the remaining six months or so of her life. The male must therefore take measures to ensure that no other males will inject her with more sperm that might displace his. During mating he anoints her with an anti-aphrodisiac, a smell that deters any other male from mating with her.

Other butterflies have other methods of preventing subse-quent copulations. After delivering their sperm, they inject a plug of a malleable substance that rapidly hardens on contact with the air to form a kind of chastity belt, so large, so awkward and so impenetrable that no other male can by-pass it. Male mosquitoes and fruit flies do a similar thing. Even mammals use this technique to protect their paternity. The male hedgehog, after ejaculating his sperm, produces a kind of gum which seals the female's orifice. Rats, bats and some marsupials do the same.

Dogs have a slightly different way of achieving the same end. It is known to every dog breeder, but its function is often not recognised and considered some kind of unfortunate accident. After copulation, the male dog dismounts by removing his forelegs from around the female's back and replacing them on the ground. But his penis still remains within her, so the pair are tied together. This is because, just before the ejaculation of the sperm, the base of his penis swelled into a bulb. Unable to withdraw, whether he wants to or not, he now lifts one of his hind legs over the female's back so that the pair, still fastened together, are tail to tail and facing in opposite directions. They may remain so for about half an hour or more. Eventually the male's swelling subsides and the two can separate. By this time, his sperm has reached her eggs within her oviduct and has fertilised them. Now, even if she were to mate again, these eggs at least will develop into his pups.

Male dragonflies have yet another system of giving their sperm priority. The male's method of copulation, like that of spiders, is not straightforward. He produces his sperm in the normal way from a pore at the end of his body, but he then arches his abdomen forward and transfers it into a special copulatory device on his underside near his thorax. This is called a penis though it is, in fact, his secondary and not his primary sexual apparatus. When he meets a female, he seizes her by the back of her neck with the claspers at the end of his abdomen. She then curves her body downwards and forwards until it touches his penis. The two are now connected in a wheel and they may remain so for as long as an hour. Most matings last about twenty minutes. Throughout this time, the male's penis is within the female, but for the first nineteen minutes or so, no sperm is being passed. The tip of his penis is armed with a variety of barbs and hooks and, as it moves within the female, it effectively routs out any sperm that a previous male may have deposited there. In some species, the tip of the penis now inflates, ramming the earlier sperm into the far recesses of the female's reproductive tract where she cannot utilise it. In others, the penis tip carries a flange beneath which the predecessor's sperm is trapped. Only after all this, in the last minute of a twenty-minute copulation, does the male dragonfly inject sperm of his own.

So every animal, male and female, strives with a wide variety of stratagems to ensure that his or her genes, and not those of a rival, will combine with those of the best possible mate and be passed on to the next generation. Naturalists tend to assume that if they witness an individual behaving in a particular way, all others of that species will act similarly. Yet again and again, as our knowledge increases, animals prove to be more variable and more inventive than we may suppose. Starlings in northern Europe migrate while those in Britain are mostly permanent residents; lions hunt in one way on the grass plains of the Serengeti but use a quite different method in the Kalahari Desert. And when it comes to the most crucial phase of their lives, reproduction, animals may alter their behaviour to suit their environmental and social circumstances. Take, for example, such a common species as the hedge sparrow or dunnock, so abundant in British suburbia. Naturalists have always known that the female dunnock builds her nest and incubates her eggs unaided and that the male does nothing more than assist in feeding the young. The home-life of the dunnock seemed, on the face of it, to be that of a staid monogamous pair, even if the male was somewhat negligent as a parent. Only when a whole population of them was fitted with leg-rings so that individuals could be identified, did ornithologists realise how socially enterprising dunnocks could be.

English gardens vary considerably in the amount of food and shelter they offer to a bird. Some, with wide lawns and pavings are poor; others, thick with shrubberies and flower beds, are full of food. Female dunnocks claim territories based on the amount of food they can provide. A rich garden may be shared between several females, whereas one of the same size but consisting largely of lawn may be able to support only one. Males, on the other hand, will claim as big an area as they can manage to defend against rivals by singing or by physical confrontation. If the territories of a male and female approximately coincide, then the pair will indeed be monogamous, the male dutifully helping the female to feed their nestlings with insects. Such a pair will raise on average a brood of five nestlings.

If the garden is particularly rich in food, however, then two females may be nesting within the territory of a single male. He will not allow another male within that range, so he finds himself with two mates. Each builds a nest, he copulates with both and brings food to both broods. However, he only has one pair of wings. Hard though he labours, he cannot provide both families with as much food as he would bring to a single one. So the size of each brood is somewhat smaller than that of a monogamous pair. Each of his females raises, not five, but only four or even three nestlings. He, however, has benefited from the situation, for he has fathered between seven and eight offspring.

On the other hand, where there are lots of lawns and little cover, a single female will find several males singing lustily and claiming different parts of her territory. So although she only builds one nest, she may accept two mates. One of these, after a series of contests with his rival, becomes dominant and ostensibly her partner. He copulates frequently and conspicuously with her. You might think that the subsidiary male faced with such a situation would go and try his luck elsewhere. But by now most territories are occupied and it is to the female's advantage to have more than one male bringing food to her chicks. So she seeks out the subordinate male in the shrubbery and there, quietly and with the minimum of fuss, mates with him too. Thus encouraged, he remains and helps to bring food to the chicks of which the dominant male appears to be father. But each egg is the result of a separate copulation, so who is to tell? The trio working together may succeed in raising seven or eight young. Most may well be the progeny of the dominant male and some may be the offspring of the subordinate. But the female has done better than either. All of them carry her genes.

Nor is this the complete list of variations in the sexual partnerships made by dunnocks. Sometimes two males will share two females, each mating with two partners. In other circumstances, two males will share three females. The inventive dunnocks are able to modify their behaviour to ensure that they produce the maximum number of young that their particular territory can support.

Human beings, in their egocentric way, tend to suppose that the marital arrangement they happen to practise themselves is the norm. For many people, that means a monogamous pair who stay together throughout their lives and so are able to help one another in bringing a succession of young to independence. Very few other mammals arrange their affairs that way. Even among birds which because of the requirements of the young, are mostly monogamous, a partnership that lasts for life is unusual. But it does occur.

Two species of great albatross live in the southern hemisphere, the wanderer and the royal. They are the biggest and the most long-lived of all flying birds, with a wing-span of over three metres and a life-span of fifty years or more. An individual spends the years of its immaturity continuously at sea, where it feeds by plucking squid, krill and fish from the surface of the water. When it is about five years old, it finds its way to a breeding colony, usually the one in which it was hatched. There it meets other youngsters of its own age and starts a series of courtship displays that are the longest of all bird dances. They bill, one bird rubbing its beak back and forth over the tip of the other. They clapper, chopping their mandibles together to make a noise like a football fan's rattle. They sky-point, lifting their beak vertically and making a moo-ing call. And in their most spectacular demonstration of all, they stretch out their immense wings and dance ponderously around one another. These movements are linked in long sequences that occupy them for hour after hour, day after day, for weeks on end. Usually they form groups of half a dozen or so, with particular pairs dancing regularly together, but sometimes, if there is a pause in the performance, a bystander will butt in and take over, as though it were recognised that this was an excuse-me dance.

At the end of the season, when breeding pairs have already laid their eggs, the adolescents will leave the colony and go back to sea and resume their separate foraging journeys. But next year when they return, they will take up where they left off. Associations inaugurated the previous year may be strengthened. Even now, their displays do not necessarily lead to mating. Indeed, such pairs may dance together for two or three seasons, becoming especially attached to each other, before ultimately they copulate and together construct the bowl of mud and vegetation that constitutes an albatross nest. Into this the female eventually deposits a single huge egg.

Throughout their first breeding season, the pair react affectionately towards each other, repeating to a somewhat lesser degree the performances in which they indulged when they first met. Incubation, however, is a particularly demanding business. It may take up to eighty-five days, longer than that of any other flying bird and in the cold islands of the sub-Antarctic where many of the great albatross nest, the egg would quickly chill if it were left for more than a few moments. The pair take turns at this chore. While one sits the other goes to sea to feed. It may travel as much as two thousand kilometres across the southern ocean while its mate remains steadfastly protecting the egg. Several weeks may well pass before the forager returns to take its turn and allow its partner to go away and feed.

When the chick at last hatches, the adults' labours intensify. Every day one of them flies off to collect food, digesting it while still at sea and regurgitating it for the chick as a rich concentrated oil, a method that enables a parent to bring back the maximum amount of nutritional calories. After three weeks, the rate of feeding diminishes and the parents travel farther and farther to collect food. Tracking such adults by satellite has revealed that, almost unbelievably, they may fly over eight hundred kilometres in a day, skilfully exploiting the wind and gliding for long periods, scarcely flapping their wings. But the strength of their pair bond is such that, even after a journey of several thousand kilometres, they will still come all the way back with full stomachs to feed their chick and allow their partner to leave the nest and feed.

The young bird takes a very long time to grow up. It has to develop strong wings, for once it leaves the nest it must stay in the air for a very long time and be away from land for years. Building adequate muscles and bones to enable them to do this requires a lot of food. The adults work hard ferrying food but, even so, ten months pass after the egg was laid before the youngster spreads its wings and glides away across the ocean for the first time.

Having successfully reared their chick, the parents can now look after themselves. They need more than two months in which to regain their breeding condition and they go away to sea for over a year. Then two years after they last laid, each having followed its own way, they return to meet one another again on the same nest site or close by it. The long labours necessary to rear their offspring dictated that male and female should remain together if they are to be successful as parents, and the pair bond developed by the dances of their betrothal seems unbreakable.

Anyone who spends any time watching animals has to conclude that the overriding purpose of an individual's existence is to pass on some part of itself to the next generation. Most do so directly. A few, such as members of the dwarf mongoose team, worker bees or the scrub jays who help at the nest, do so indirectly by assisting a breeding individual whose genes they share. Inasmuch as the legacy that human beings bequeath to the next generation is not only genetic but, to a unique degree, cultural, that is true of us too.

To achieve this end, animals, including ourselves, endure all kinds of hardships and overcome all kinds of difficulties. Predators are foiled, food is gathered, rivals are fought, mates selected and the complexities of copulation negotiated until at last the next generation is brought into existence. Then it is their turn to carry the genes through yet another cycle of the never-ending trials of life.

Acknowledgements

I doubt if any one individual could have witnessed first-hand all the many activities described in the preceding pages. Certainly I have not. In many instances, I have based my accounts on detailed descriptions published in learned journals and elsewhere by zoologists. These sources are so numerous and disparate that listing them all would hardly be appropriate in a book of this kind, but like anyone writing on scientific subjects, whether specialised or popular, I could only write what I have because of the labours and observations of innumerable predecessors.

Some of the animal dramas I have of course seen first hand, but I would have seen few even of these were it not for the help and guidance generously given by researchers who have studied the animals concerned, often for many years. I became personally indebted during our journeys in Africa to Christophe and Hedwige Boesch, Mark Collins, Hussein Isack, Paul Kabochi, Cynthia Moss, Craig Packer, Joyce Poole and Rudiger Wehner; in North America to Victor van Ballenberghe, John Fitzpatrick, John McCosker, Gary McCracken, Chris O'Toole, Henriette Richard, Mel Sunquist and Glen Wolfenden; in South and

Central America to Anne Brooke, Claudio Campagna, Nigel Franks and Larry Gilbert; on Christmas Island to Hugh Yorkston; in the Bahamas to Denise Herzing; in Samoa to Lui Bell and Karl Marshall; in Malaysia to Ivan Polunin and Jason Weintraub; in Russia to Alexei Suvarov and Kathy Wynne-Edwards; in Ireland to Christopher Moriarty; in New Zealand to Chris Robertson; and in Australia to Dawn and Cliff Frith, Chris Hill and Peter Jacklyn.

In some instances, I have used as a source of information film shot by cameramen working on the television series. Their sharp, discriminating and knowledgeable eyes revealed many aspects of animal behaviour with a special clarity and on occasion some details that had never been noted before. So I am very grateful to them too. Nor are cameramen the only members of the television team to whom I am indebted. Researchers discovered stories that hitherto were unknown to me; recordists and production assistants, listening to my words being spoken in the field, gently reproved me if my thoughts were being expressed unclearly; directors and producers, by hard argument, forced me to clarify my ideas and often converted me to their way of thinking.

I am also most grateful to Myles Archibald of William Collins Publishers who suggested that there should be a new edition of the book, and to Rachelle Morris and Laura Sutherland who brought together the photographs with which it is illustrated.

Finally I must thank Professor Matthew Cobb of the University of Manchester who has read the entire book to ensure that, although there have been many advances in our understanding of animal behaviour in recent years, the text still retains its accuracy.

Picture Credits

Index

The names of the animals used in the text are here accompanied by their scientific names, where these differ substantially. In many cases, these Latin names identify the animal more precisely than do the English names.